WHY SHARKS MATTER

WHY SHARKS
MATTER

A Deep Dive
with the
World's Most
Misunderstood
Predator

DAVID SHIFFMAN

Johns Hopkins University Press
Baltimore

Johns Hopkins University Press
2715 North Charles Street
Baltimore, Maryland 21218-4363
www.press.jhu.edu

Library of Congress Cataloging-in-Publication Data

Names: Shiffman, David, author.
Title: Why sharks matter : a deep dive with the world's most misunderstood predator
    / David Shiffman.
Description: Baltimore : Johns Hopkins University Press, 2022. | Includes index.
Identifiers: LCCN 2021022868 | ISBN 9781421443645 (hardcover) | ISBN
    9781421443652 (ebook)
Subjects: LCSH: Sharks. | Sharks—Behavior. | Sharks—Conservation.
Classification: LCC QL638.9 .S525 2022 | DDC 597.3—dc23
LC record available at https://lccn.loc.gov/2021022868

A catalog record for this book is available from the British Library.

*Special discounts are available for bulk purchases of this book. For more information,
please contact Special Sales at specialsales@jh.edu.*

To my family, for always supporting my dream of becoming a marine biologist

Contents

Color plates follow page 150

WHY SHARKS MATTER

Introduction

In the end, we will conserve only what we love; we will love only what we understand; and we will understand only what we are taught.

—Senegalese environmentalist Baba Dioum

Thanks for reading this book and letting me be your guide to some of the most fascinating, important, threatened, and misunderstood animals on the planet.

My name is David, and I'm a marine conservation biologist who studies sharks and how to protect them. You may know me from social media,* where I am always happy to answer anyone's questions about sharks. This seems only fair, since I'm one of the lucky few who gets to live my childhood dream every day.

I've been captivated by sharks as long as my family can remember, and I've known for essentially my whole life that I wanted to be a marine biologist who studies them. While most kids go through either a shark thing or a dinosaur thing, I went through both phases. In the end, though, I chose sharks. My parents have always supported this crazy ambition of mine, but I suspect they believed that I would grow out of it. Clearly, I never did.

You might guess that an ocean lover like me grew up right on the beach, but I was raised outside of Pittsburgh, Pennsylvania. (According to DistanceBetweenCities.net, my childhood home was 368 miles from the nearest ocean.) So how does someone who grows up landlocked become obsessed with the sea? I read every book I could get my hands

*Follow me on Facebook, Twitter, or Instagram @WhySharksMatter.

on, and I watched every documentary I could find. We were members of the Pittsburgh Zoo growing up, and I loved visiting the shark tank at what's now their PPG Aquarium complex.

I got scuba certified as soon as I was old enough and convinced my family that many of our annual vacations should include scuba diving. My younger brother, Eric, was hesitant to join me at first, but I eventually convinced him that most ocean animals are pretty small and generally mind their own business. We both needed to change our bathing suits, however, after his first open water dive, when an extremely large loggerhead sea turtle rushed at us from under a ledge.

Dad loves diving, while Mom is a good sport, content to read a book by the hotel pool no matter what country we're in. I've been diving on five continents so far and have a little over 800 logged dives. (I should briefly mention that zero of these dives were for work. A fun fact is that you can absolutely be a marine biologist if you are medically unable to scuba dive—I know lots of marine biologists who don't even go on boats because they get seasick.)

Like all the cool kids, I went to marine biology summer camp in middle and high school. I spent five summers attending Seacamp on Big Pine Key, Florida, where I took one of my first formal marine biology classes. There, I learned about ichthyology (the study of fishes), coral reef ecology, and animal behavior. I even completed a course on the biology and behavior of sharks. The science instructors at Seacamp are all trained marine biologists, and meeting such amazing people only reinforced that this was the career I wanted.

While attending Seacamp, I met my first professional shark scientist, Dr. Jeffrey Carrier. Jeff created the shark biology course at Seacamp years earlier and still uses the camp as a staging ground for some of his research on nurse sharks in the Florida Keys. Jeff is one of the many passionate, brilliant, and colorful characters you'll meet in this book. In 2020, I fulfilled a long-cherished dream of publishing a peer-reviewed scientific journal article with Jeff. I was even the lead author!

When it came time to apply for college, I finally had a chance to live near the ocean. I eventually chose Duke University, which has a coastal marine lab campus in North Carolina's Outer Banks. I ended up spending a semester and a summer at the marine lab, where I studied the feeding ecology of smooth butterfly rays, a fairly absurd looking relative of the stingray. I also studied abroad at James Cook University in Townsville, Australia, right on the Great Barrier Reef.

My college application essay was, of course, about sharks, specifically about going scuba diving in the shark tank at the New Jersey State Aquarium (now the Adventure Aquarium) with my father. I opened my essay with the line "Don't worry, Dad, they don't usually eat people." This was particularly apropos because the tank's residents included large sand tiger sharks, which South Africans call ragged-tooth sharks because of their large and nasty-looking teeth. I included a photo of me next to an extremely large sand tiger shark. Years later, I was speaking at a Duke-sponsored event for prospective applicants, and a student asked one of the admissions officers on the panel what their all-time favorite essay was. She responded by praising a piece about swimming with sharks, despite not knowing that the intrepid author was sitting right on stage with her. I still maintain that essay got me into Duke.

After graduating from Duke, I returned to Seacamp, where I worked as a camp counselor and marine biology instructor. I even got to teach the shark biology class that had so inspired me as a middle school student. I've since learned that one of the students I taught shark biology to at Seacamp returned years later as the shark biology instructor. In between working at Seacamp and starting graduate school, I took a course at the Bimini Biological Field Station, aka Shark Lab, taught by Daniel Abel and Dean Grubbs. Their book *Shark Biology and Conservation* notes that it is based on that very class.

I earned my master's degree in marine biology at the College of Charleston, where I studied the feeding ecology of sandbar sharks. I worked with Dr. Gorka Sancho from the College of Charleston and Bryan Frazier (who you'll meet later in this book) from the South Carolina Department of Natural Resources. This is part of the reason I love sandbar sharks. Follow #BestShark on Twitter and Instagram and you'll find that I spent years talking up this underappreciated species while senior colleagues teased me for having a supposedly boring shark as my favorite (see Plate 1 in the color insert for a closer look at this beautiful

Holding a young-of-year tiger shark in the Bimini Shark Lab holding pond. Sadly, I no longer have that amazing coral reefscape bathing suit. *Courtesy of the author*

shark).* I got to see more sharks during my three years at Charleston than in the rest of the years of my life combined (so far). This is because the shark survey I participated in focused on smaller species and the coastal nursery areas they use. Studying lots of small sharks rather than a few larger sharks meant that it wasn't unusual for us to see a few hundred sharks in a day.

I also worked as a volunteer diver in the South Carolina Aquarium's shark tank during my time in Charleston, helping to maintain the exhibit and feed the fishies. I wasn't ever bitten by a shark in that tank, but

---

*Why do I believe that sandbar sharks are the best sharks? They're one of the most studied species of sharks in the world and have contributed much to our understanding of marine life. They do well in captivity, which means they are the first shark millions of kids ever see— and I am living proof that seeing sharks in a tank can inspire a lifelong love of the ocean. And they've made all these contributions to research, management, and public outreach without having fancy tails like a thresher, fancy heads like a hammerhead, or fancy stripes like a tiger shark. Finally, I love them because they were the first shark species I ever spent a lot of time with, and because they are unsung heroes, given all they've done for us.

there was a particularly nasty red drum in there. And divers had to wear hoods not because the water was cold but because the sea turtle tended to confuse ponytails with seaweed.

I began to zero in on shark conservation at University of Miami while working on my PhD in interdisciplinary environmental science and policy. There, I studied under Dr. Neil Hammerschlag. I also learned from experts in turning science into science-based environmental policy, and I wrote a review and guide to all the available policies to protect and manage sharks that forms the basis of a good chunk of this book. I got to help train the lab's interns, reinforcing a love for teaching and mentorship that was first sparked while I worked at Seacamp. Finally, I got to spend many days on the water tagging sharks (see Plate 2 in the color insert for a picture of me holding a shark during the tagging process). A great thing about living in South Florida is that sharks are there year round, unlike South Carolina, where many species leave in the fall and winter. During this time, I got to work with some amazing species of sharks, including great and scalloped hammerheads (see Plate 3 in the color insert) and even the #BestShark itself.

My postdoctoral fellowship at Simon Fraser University outside of Vancouver was through a program called the Liber Ero fellowship in conservation biology that focused on interdisciplinary policy-relevant environmental science. My supervisor in the Liber Ero program was Dr. Nick Dulvy, who at the time was the co-chair of the IUCN Red List of Threatened Species' shark specialist group. (The IUCN is the International Union for Conservation of Nature, and you'll learn about them later in this book.) At Simon Fraser, my work focused on North American shark management practices and public perceptions of shark conservation issues.

I am currently a postdoctoral researcher at Arizona State University's Washington, DC, center, an institute that focuses on studying science policy and environmental policy. I teach an online marine biology course for ASU.

My career has taken me on some amazing adventures. I've met some fantastic people, and I've learned a lot about sharks and how to protect them. Through it all, I've developed a passion for public science engagement. I have spoken to thousands of people around the world (including school groups in all 50 states and 19 countries) about sharks and marine biology. I answer thousands of people's questions about sharks in weekly "Ask Me Anything" sessions on social media. I've written about sharks for outlets like the *Washington Post*, *Scientific American*, *Gizmodo*, and *National Geographic* and in a monthly column in *Scuba Diving* magazine. To date, I've done more than 300 interviews with the popular press. I still love sharks just as much as I did when I was a kid, but after fifteen years of studying them and their conservation, I now know a lot more about how we can protect them and why we should do so. The goal of this book is to try and pass on some of that knowledge.

Structure of the Book

> At any given moment there is a shark behind you. It might be a thousand miles away, but there is a shark behind you.
>
> > And that shark is totally behind you, it is supporting you in whatever you do and wants you to succeed.
>
> —A viral Tumblr conversation between users the-rain-monster and friedcherryblossomprincess

In this book, I hope to teach you why sharks are remarkable and awe-inspiring animals, why we're better off with sharks than we are without them, and what you can do to help protect the alarming and increasing number of shark species of conservation concern. For those of you who are already convinced that sharks are awesome and need to be protected, I'll explain the many possible conservation and management policies that can be used to protect sharks, including many expert-supported, evidence-based policies that don't seem to be discussed much among non-expert enthusiasts.

In this book, you're going to dive deep into the science of shark conservation. You'll be introduced to the latest science showing that sharks provide a myriad of incredible ecological benefits to the ocean (and associated economic benefits to humans). You'll learn all about the many threats that sharks face and creative solutions to solving the problems those threats pose. I'll show you how to find reliable information and how to ignore unreliable information, how to identify credible experts and organizations, and how you personally can help sharks. And of course, I'll be treating you to a plethora of fascinating facts about my favorite denizens of our vast oceans.

Throughout it all, you'll meet some of the brilliant and passionate people from the world of shark research, conservation, and management. It's been a privilege to know them, and it's an honor to introduce them to you in these pages. You'll also hear all about my sometimes wild experiences as a shark scientist and my research projects. Although I'm taking care to fairly present all sides of some ongoing disputes about the best ways forward, fair warning, you're definitely going to hear my personal thoughts on who seems to be right.

Don't worry about taking notes or looking up the papers and news articles that I reference throughout this book. In addition to listing them chapter by chapter in the book's bibliography, I've organized links to them all for you on the book's website at https://jhupbooks.press.jhu .edu/title/why-sharks-matter. The website also contains links to videos or photographs demonstrating some of the biology and behavior of sharks I describe here, along with ways for you to learn more about the scientists and environmentalists I feature in this book, including their social media handles and links for you to donate in support of their important work.

I hope that you learn something, and that you enjoy the journey while you do. Please direct any questions and/or hate mail to WhySharksMatter @gmail.com, or find me on social media @WhySharksMatter on Instagram, Facebook, and Twitter.

1 » Shark Basics, and Fun Facts to Keep You Reading

What Is a Shark?

> Sharks have been living for over 400 million years. They came through the great mass extinctions, and competed favorably with the rapidly evolving bony fishes, and today they are among the top predators of the sea. During that time, sharks have been honed to evolutionary perfection, with a body that yacht designers can only envy, with wide-ranging sensory systems that military technologists would die for.
>
> —From Steve and Jane Parker's introduction to *The Encyclopedia of Sharks*

I'm sure that you're eager to get started, but before we dive in and turn you into an effective ally for the science-based shark conservation movement, we need to make sure that everyone is on the same page and understands the foundational material. That's right, it's time for everyone's favorite part of a new educational adventure: the basics, in which we start at first principles and define our key terms.

. . . Wait, no, come back, keep reading! This stuff matters a lot, I promise. Much of the later content in this book will make sense only if you understand the background material. Even if you think you know everything there is to know about shark biology and behavior, I'd encourage you to check out this part. As a reward for your due diligence, I'll sprinkle each section with some fun, little-known facts about sharks. (If you look at Plate 4 in the color insert, you'll see a photo of me from a 2018 conference that shows that science nerds can indeed have fun. While I take my job very seriously, I do not take myself very seriously!)

Before we can talk about how to save sharks (or why you should want to or what they need to be saved from), we need to start at the start: What exactly is a shark? One of the most common questions I get when I speak with school groups is "Are sharks fish?" Some students seem to believe that sharks are mammals. A few believe them to be amphibians, and many think they're their own group at the same taxonomic level as mammals and fish. Let's settle this once and for all: sharks are indeed fishes, they are not mammals, they are not their own thing. And please note that I said "fishes," not "fish." The word *fish* can be both singular and plural: one yellowfin tuna is a fish, twenty yellowfin tuna are *fish*. But a group of multiple species are fishes: a group containing one yellowfin tuna and one skipjack tuna are *fishes*.

So yes, sharks are fishes, but they are part of a different group of fishes than the tunas and goldfish and bass that many people are more familiar with. This split occurs at the taxonomic level of class (remember your "kingdom, phylum, class, order, family, genus, species" from middle school life science?). Tunas, goldfish, bass, and most of the known species of fishes are in the class Osteichthyes, which classical language nerds will know means "bone fish" (osteo = bone; ichthyes = fish). Not surprisingly, bony fishes have a skeleton made out of bone, just like you do. In contrast, sharks and their relatives form the class Chondrichthyes, or cartilaginous fishes (chondro = cartilage), which means that their skeletons are made not out of bone but out of cartilage. Take your hand and scrunch up your ears a little: the hard structure you feel is cartilage. It's lighter and more flexible than bone. Among other things, this means that if you grab a shark by its tail, it may be flexible enough to turn around and bite your hand.

Do you recall that I mentioned "sharks and their relatives" above? Class Chondrichthyes includes not only sharks but also skates, rays, and chimeras. Chimeras are also sometimes called ratfish (which is a terrible name for such a cool group of animals) or ghost sharks (which is a friggin' awesome name). In case you've had a long-simmering desire to learn the difference between skates and rays, while both are flat (and sometimes called "flat sharks" by advocates hoping to get people who care about sharks to also care about skates and rays), the difference has to do with their tail and fin anatomy (skates have shorter, stubbier tails) and their reproductive strategies (skates lay eggs, rays give live birth). They're also in different orders.

Incidentally, not all rays are stingrays; manta rays, for example, are not. While stingray stingers are a nasty piece of biological weaponry, worry not, they can only be deployed defensively as a reaction to being stepped on—which means that the story that Steve Irwin's untimely death involved being stung "hundreds of times" by the stingray that killed him is just preposterous. Stingrays are responsible for only about 30 known human deaths in the entirety of recorded human history, but lots of injuries—many of which could easily be prevented by doing a goofy walk called the "stingray shuffle," which startles them up from the seafloor without giving you a chance to step on them and trigger their defensive reflex. Although some stingrays do have venom in their barbs, not all do.

Some readers may also be familiar with the term *Elasmobranch*, which means "plate gills" and is sometimes wrongly believed to be a synonym for shark. The class Chondrichthyes includes sharks, skates, rays, and chimeras, while the subclass Elasmobranchii includes only sharks, skates,

A 3D printed model I made of a Pacific cownose ray's stinger. Check out the nasty serrations. Though the stinger in the model is twice the size they really grow in order to make it more visible—an advantage of 3D printing—you can see how this would ruin your day if it went in your leg.
Courtesy of the author

and rays—a source of somewhat nerdy contention between chimaera biologists and the American Elasmobranch Society. While skates, rays, and chimaeras sometimes fulfill similar ecological roles and face similar conservation challenges, in this book we will focus overwhelmingly on sharks. I do this for two reasons: (1) I know much more about sharks than I know about skates, rays, or chimaeras, and (2) people tend to care more about sharks than they care about these other elasmobranchs. So, I'll meet you where your interests are while encouraging you, when the subject permits, to expand your horizons a little and learn about other animals.

Getting back to basics here, sharks are cartilaginous fishes that are members of the class Chondrichthyes. However, sharks don't all look like the great white in *Jaws*. They come in an extraordinarily diverse array of shapes, sizes, and colors, with diverse behaviors to match. So how can you tell that what you're looking at is a shark?

What Makes a Shark a Shark: Biology and Physiology

What we are dealing with here is a perfect engine, ah, an eating machine. It's really a miracle of evolution. All this machine does is swim and eat and make little sharks. And that's all.

—Matt Hooper (Richard Dreyfuss), *Jaws*

Assuming that you aren't able to examine the skeleton and determine that it's made of cartilage and not bone, how can you tell that the animal you're looking at is a shark? Thanks to *Jaws*, the most iconic feature of a shark is probably the *dorsal fin*, the fin on the animal's back that sticks out of the water when a shark gets close to the surface. However, dolphins have a dorsal fin, too. Many a marine biologist has been injured while violently rolling their eyes at fearmongering news footage of a "shark" stalking a swimmer that clearly shows the dorsal fin of a dolphin. Many shark species have two dorsal fins, including a second (usually) smaller one set farther back. Sharks also have other sets of paired fins that appear on either side of their body, including the pectoral fins (located about where you'd expect arms to be) and pelvic fins.

One way you can tell that you are looking at a shark is by examining the tail, or *caudal fin*. Unlike a whale or dolphin tail, which is horizontally oriented and moves up and down to propel the animal, shark caudal fins are vertically oriented and move side to side. They also can be *heterocercal*, which means that the top part of the fin and the bottom part of the fin are different sizes or shapes. The most extreme example of this is the thresher shark; the top half of a thresher shark's tail can be longer than the rest of the shark combined. Check out the video on this book's website that shows thresher sharks using their tails as whips to stun prey. I shrieked the first time I saw footage of this amazing behavior, and it's always a crowd-pleaser at my public talks.

Like other fishes, sharks have gills, and we can see vertical *gill slits* on either side of their heads. Like bony fishes—and unlike mammals and marine reptiles such as sea turtles, which have lungs and need to breathe air at the surface—sharks have gills, which let them breathe underwater as long as water flows over them. These organs are just a little different in external structure from bony fish gills.

This is a good opportunity for some mythbusting. A thing that "everyone knows" about sharks is that they have to keep swimming forward constantly or they'll drown. I've seen this claim show up in all kinds of places. But is it really true? Well, it certainly is for some species, which is why sharks that are accidentally caught by fishing gear are often dead by the time fishers try to release them. However, some species can hold still on the seafloor and pump water over their gills.*

In the southeastern United States, the most common and well-known species that doesn't need to swim constantly in order to breathe is the nurse shark. It has an unearned reputation for being weak or lazy (though admittedly, I did once see a nurse shark sitting under the same coral head for five straight days). Shark researchers will tell you that nurse sharks are often the most intense species that we have to restrain

---

*Most sharks have five gill slits on each side of their head. A group of sharks called sixgill sharks have six gill slits on either side of their head. There is also a group of sharks called sevengill sharks; would you care to guess how many gill slits they have on either side of their head?

and study. They engage in an alligator-like rolling behavior I've heard referred to as the "death roll of death," which generates an unbelievable amount of force. It can take half a dozen scientists to hold a large nurse shark mostly still. In fact, the worst shark-induced injury I've ever gotten was from a nurse shark. I've never been bitten by a shark, but one time an intern who was attempting to secure a nurse shark's powerful tail lost her grip, and the resulting Indiana Jones whip crack meant that I couldn't sit down for three days. I still occasionally get phantom tingling butt pain when I see videos of nurse shark workups.

Not all sharks respond to the stress of capture in the same way, and nurse sharks represent an extreme position on this continuum. While some species die pretty quickly after being caught by fishers, nurse sharks are basically swimming tanks—and not just because their tough skin can't be penetrated by some of the surgical steel research tools we use to take skin and muscle samples from other species. I once saw a nurse shark that appeared to have been repeatedly shot in the face; other than some mild scarring, he seemed totally fine. My favorite indestructible nurse shark story could also fairly be referred to as a Florida Man story. (For those who don't know, Florida Man is a generic descriptor for any particularly wacky criminal suspect in the Sunshine State. It became a popular meme when people began to notice a high volume of unusual behavior from Florida residents.) Some guy in Fort Lauderdale caught a nurse shark, put it in the back of his pickup truck, and drove it to a nearby supermarket. He then slapped it down on the sidewalk and proceeded to spend *hours* trying to sell it to people leaving the store. Eventually, police were called, and he was forced to drive the shark back to the ocean and release it. Amazingly, despite spending hours out of the water sitting in a hot Florida parking lot, it swam away seemingly uninjured.

While we're talking about sharks constantly swimming, I should also mention the role that this behavior has in generating lift, which keeps the shark from sinking to the seafloor. Another difference between sharks and the bony fishes is that sharks don't have a *swim bladder*, a gas-filled sac found in bony fishes that helps them stay afloat and keep still without bobbing up and down in the water column. Instead, sharks have huge, oily, buoyant livers (how large? See Plate 5 in the color insert) and generate lift by moving forward, with their pectoral fins functioning kind of like airplane wings. Incidentally, until we figured out how to synthesize vitamin A in labs, sharks used to be caught for their vitamin-rich liver oil.

These days, while shark liver oil is sold for a variety of purposes, including use in some vaccines as an adjuvant (that is, to enhance the body's immune response), sharks aren't often targeted just for their livers anymore. A petition went viral while I was writing this book, claiming that hundreds of thousands of sharks would be slaughtered to help make COVID-19 vaccines; it was nonsense. Some pharmaceutical companies use shark liver oil–derived squalene in their vaccines, but there's plenty available from sharks who have already been killed by fishers, and there's no evidence to suggest that hundreds of thousands of additional sharks would need to be killed to supply this demand.

Let's go back to what makes a shark a shark. Another unique characteristic of sharks has to do with their skin. Dolphin skin is smooth, bony fishes' skin is covered in tiny plates called scales, but what about shark skin? Sharks have a fascinating arrangement of tooth-like structures called *dermal denticles* which point backward, snoot to tail. The composition and arrangement of these denticles serves a practical purpose: they help make sharks more hydrodynamic, cutting down drag as they move through the water. Some human swimmers even use suits that were designed based on this same principle.

This backward arrangement also means that, if you pet a shark in one direction, they feel quite smooth. If you pet it in the other direction, though, the denticles are sharp enough to cut your hand or tear field research clothing.* Most shark researchers I know have had some sort

---

*Shark skin texture is the source of a running joke in my favorite science meme group, Wild Green Memes for Ecological Fiends. A webcomic by Branson Reese entitled "The Person Who Discovered Sharks" included the text "Smooth lions are attacking me." Reese later remarked that he had accomplished his lifelong dream of getting hundreds of nerds to angrily explain the real texture of shark skin to him. Reese kept this joke going for a long time, even inserting fake text in a textbook that claimed that sharks are "Smooth as hell in all directions." Through a preposterous serious of coincidences that only seem to happen in my ridiculous life, the comic inspired a song that features me as a guest vocalist. Amazingly, the song and associated meme (go team "At Least 74 Practicing Shark Researchers"!) raised thousands of dollars for rainforest conservation.

of negative experience with shark skin, which can cause something we call "shark burn." One time, I looked down at a lemon shark we were working up and noticed that it had a clump of hair on it. "That's weird," I thought, because sharks don't have hair. Then I realized that the shark had twitched suddenly, resulting in its dermal denticles scraping across my leg, taking the hair and a little bit of my skin right off. The noise I emitted next was extremely masculine, no matter what the interns on the boat that day may claim.

Finally, what about shark teeth? Like sharks themselves, these teeth come in a variety of shapes, sizes, and designs. Some are especially well-suited for a specific purpose. The mako shark's long, pointy teeth, for example, are great for piercing prey. The tiger shark's can opener–like teeth can crack open a sea turtle shell. Nurse sharks have ray-like crushing plates that are great for eating shelled prey. As you can imagine, teeth can not only help non-experts tell a shark from a bony fish, they can tell you what specific species of shark the teeth came from. (And if the tooth is black, as you often see in those very tacky shark tooth necklaces that are free with purchase at beach gear stores, then it's fossilized, which means that the shark died before there were humans.)

A cool thing about shark teeth is that sharks never stop growing them. Humans have baby teeth and adult teeth and that's it; sharks are constantly regrowing new teeth their entire lives. If you look at a close-up of a shark's jaw, you can see they're built like a conveyor belt, with another row of teeth ready to move into position and a smaller row behind that in the process of being grown. See this book's website for an incredible photo of a shark's jaw and the tooth conveyor belt.

Shark Biodiversity and Biogeography

According to the latest edition of the field guide *Sharks of the World*, there are 536 recognized species of sharks. They range in size from the dwarf lanternshark, which could fit in your hand, to the school bus–sized whale shark. Many—like the sandbar shark (#BestShark)—have the particularly sharky shape you're familiar with from movies or from visiting your local aquarium, but some, like the angel shark, are flat and capable of burying themselves in the sand to wait for prey. Some deep-sea weirdos like the frilled shark are almost snake-like in appearance and movement. Many are gray or brown in color; some are blue; some, like the goblin shark, can be bubblegum pink. Some sharks have beautifully elaborate patterns of stripes or spots. Some are sleek, like the shortfin mako shark, which is among the fastest animals in the world. Others, like the angular roughshark, have just about the least hydrodynamic shape I can imagine: they look like the ocean's overinflated footballs.

Recognizable sharks have been swimming in the ocean for more than 400 million years. This means that the first shark was on Earth not only well before dinosaurs trod the land but before trees existed. Though we've lost many species over the eons, sharks as a group have survived every mass extinction event in Earth's history—which makes the conservation challenges they've faced in the past 50 years all the more heartbreaking. While we're talking about ancient sharks, let me assure you that, no, the giant and ancient megalodon is not still alive. It is definitely super-duper extinct. People claiming otherwise are lying to you, for reasons that remain unclear to me despite a decade of refuting this really strange folk legend. I've received death threats from people who believe I am part of a global conspiracy to hide the truth about megalodon. Once I even interacted with someone online who emphatically made the bizarre and obviously false claim that she had seen the US government rounding up and killing megalodons—and that she had barely escaped with her life once the shark-killing soldiers spotted her.

Sharks' habitats are as diverse as the animals themselves. Some sharks are found on coral reefs, while others, like the Greenland shark, are found under Arctic ice. (Fun fact about Greenland sharks: they have been found with digested polar bear and reindeer meat in their stomachs. These are probably the remains of scavenging animals that drowned, but I enjoy imagining a polar bear getting slurped from below as it swims between ice floes.) Some sharks live in the open ocean, where they'll never see a hard surface their entire lives. Some sharks live in the deep sea, where it's so dark that sunlight never reaches. The megamouth shark, a deep-sea animal with the world's coolest scientific name—*Megachasma pelagios*, which means "the giant mouth of the deep"—has bioluminescent gums that entice prey to swim right into its mouth.

US Navy Seals jokingly say that you can test whether there are sharks nearby by dipping your finger in the water and tasting it—if it's salty, there are probably sharks around. While technically accurate—there are sharks just about everywhere there's ocean—the implication is incomplete, because there are also sharks that live in fresh water. No, I'm not just referring to the bull shark, which Discovery's Shark Week programming wrongly claims year after year is the only shark that can enter fresh water. I'm also talking about *Glyphis* sharks, sometimes known as river sharks, which live almost their entire lives in fresh water. Unfortunately, river sharks are some of the most critically endangered sharks in the world, in no small part because they live closer to humans than ocean-dwelling sharks do.

What we already know about shark biodiversity is amazing, but it's what we don't know yet that many attendees at my public talks find shocking. We are still discovering new species all the time. A new species of chondrichthyan fish is discovered about every two weeks. Some of them get tons of media attention, like a new species of "walking shark," so called because they can crawl on their fins out of water for short periods of time, or a new species of dogfish named after shark science legend Genie Clark (Genie's dogfish, *Squalus clarkae*). Others are little known outside of science nerd circles. There's plenty left to discover. (But no, that doesn't mean that megalodon is still hiding out there.)

Unfortunately, the threats these species face are as diverse as their habitats and color patterns, which means that there's no one-size-fits-all solution. For instance, creating no-fishing zones is less helpful to a species that moves around a lot and spends limited time in protected areas. Nor is a ban on selling shark fins especially useful for the many species killed for reasons having nothing to do with their fins. Generally speaking, any solution to a complex worldwide conservation problem simple enough to fit on a bumper sticker is perhaps too simple to be helpful.

Shark Senses: How Sharks Perceive Their World

Sharks possess an impressive array of senses that they use to navigate through the underwater world and to find prey. They have all the same five senses that people have, plus two more.

Sharks' sense of smell is legendary. In some species, nearly one-quarter of their entire brain is devoted to processing scents in their watery home. Some older books even refer to sharks as "swimming noses." No, that doesn't mean that they can smell blood from a mile away or that they can detect a drop of blood in an Olympic-sized swimming pool, two common myths I've encountered. In fact, whenever you smell something, it's because tiny particles of whatever you're smelling are interacting with chemoreceptors inside your nose—think of that the next time you smell something gross! It's worth pointing out, too, that sharks aren't just smelling blood when tracking prey, but also various chemicals associated with injured or dead prey animals. And while you probably shouldn't go in the ocean if you're actively bleeding for a lot of reasons, human blood does not smell the same as fish blood to a shark.

Sharks can, however, hear the sounds of a fish struggling or something mechanical, like the sound of a boat motor, from nearly a mile away. Their *lateral line*, a mechanosensory organ that can detect vibrations in the water, is sensitive enough to feel the difference between a healthy fish swimming strongly or a sick and injured fish for which escape would be difficult.

Sharks have a reputation for poor vision; this assumption is often erroneously used to explain why they may accidentally bite humans. The reality is that, while the huge diversity in shark species means that eyesight quality depends heavily on which species you're talking about, some sharks have fairly advanced vision—it's just not designed to tell the difference between a seal and, say, a human wearing a seal blubber–like wetsuit. Sharks, after all, are extraordinarily well-adapted to the oceans that have been their homes for tens of millions of years, and it's perhaps not fair to blame them for being imperfectly adapted for conditions that have appeared in an evolutionary blink of an eye. Sharks can see better than humans in just about every underwater environmental condition, which is to say, just because you can't see a shark doesn't mean that a shark can't see you.

In my opinion, the most amazing sense that sharks and their relatives have is the electric sense; that is, their ability to detect electromagnetic fields. Platypuses also have this sense (because what's one more random addition to an animal that lays eggs and produces milk and can therefore make its own custard?). Some other bony fishes here and there also have an electric sense, but it's relatively uncommon outside of the chondrichthyan lineage. This sense is housed in jelly-filled pores on a shark's snoot called *Ampullae of Lorenzini*, which look almost like the chin stubble of a five o'clock shadow. (If you ever get the chance to witness a shark necropsy—basically the equivalent of an autopsy on humans—you'll see that a tiny bit of gooey gel comes out when the face is squeezed. Please do not try this on a living shark for what I hope are obvious reasons.)

You may wonder what use it is to have an electric sense. Sometimes prey animals hide under the sand where a predator wouldn't be able to see, smell, or hear them. Sharks, however, can still detect their presence by sensing their body systems' electricity, and can dig them up for a meal. This electromagnetic sense is also helpful when it comes to long-distance open-ocean navigation. Not only are there no street signs in the middle of the Pacific, there aren't any landmarks. Sharks can navigate the featureless open ocean by detecting the Earth's magnetic field, which lets them end up exactly in the right spot after a long migration.

Shark Reproduction and Life History

"The Whale Shark, *Rhincodon typus*, Is a Livebearer: 300 Embryos Found in One 'Megamamma' Supreme"

—Title of a 1996 paper on whale shark reproduction (this is my all-time favorite scientific paper title, narrowly beating out "Fantastic Yeasts and Where to Find Them")

You're doing great so far with the introductory material. Here's another fun fact to keep you paying attention: ichthyologist David Starr Jordan was the first scientist to describe the goblin shark. This shark is known for not only its bizarre coloration and long, pointed snout that's been described as "floppier than you'd expect," but also for having the ability to hyperextend its jaws. Jordan's life was the subject of Lulu Miller's 2020 book, *Why Fish Don't Exist: A Story of Loss, Love, and the Hidden Order of Life*. While serving as the first president of Stanford University, Jordan allegedly played a role in the 1905 murder of the university's cofounder, Jane Stanford; evidence suggests that he also helped to cover up her death. I'm still mad that this nugget of information got cut from an article I wrote for *Sport Diver* magazine about goblin sharks, so you can enjoy it here. (Jordan was also an avowed eugenicist; recently, a California elementary school that had been named after him was renamed.)

Anyway, back to sharks. The last thing I need you to understand before we can start talking about the threats sharks face is that their unusual life history makes them inherently vulnerable to overfishing pressures. Sharks have what's called a *K-selected life history*, which means that they have relatively few babies, have them relatively infrequently, and have them relatively late in life. This strategy has huge benefits; any given baby shark is more likely to survive to adulthood compared with, say, a larval tuna because the mother is able to invest more energy into each individual baby, mainly while the baby is still inside the mother. (Baby sharks grow up inside the body of the mother, either all the way to birth or for at least a little while.) Indeed, young-of-year sharks are able to fend for themselves from the moment they're born. This strategy has its disadvantages, especially in the age of human industrial-scale fishing: if many sharks are killed, it takes a lot longer for their population numbers to bounce back. Let's unpack this a little bit more.

First of all, unlike most bony fishes, sharks do not spawn. *Spawning* is when a large number of fish of the same species get together and release a giant cloud of sperm and eggs into the water. These eggs fertilize outside of the mother's body (usually, but not always, the baby fish will spend a small part of their life as plankton). Even though most of these baby fish don't survive—it's hard out there if you're less than an inch long—the spawning process can produce a huge new generation. Sharks, in contrast, reproduce by means of internal fertilization. The shark mating process is fairly recognizable, even if you're seeing it for the first time. Male sharks have two sex organs called *claspers*, which are inserted (only one at a time) into the female's *cloaca* for sperm transfer (a cloaca is a cavity at the end of the digestive tract—birds and reptiles, among others, have this as well). When my PhD lab had high school students join us for a day of shark research, sometimes they would say "I think this is a boy but I am not sure," and we would reply "then it's not a boy, if it was you could definitely tell."

Shark reproduction, by the way, is extraordinarily violent and involves lots of biting, which allows sharks to stay together while moving through a three-dimensional environment. Female sharks' skin is often twice as thick as that of males of the same species, allowing them to withstand the trauma of the bites. This also explains why we scientists sometimes laugh when you ask us if our research hurts the sharks—because normal day-to-day life clearly hurts them much more than anything we're doing.

What happens next is incredibly diverse by biological standards. Some sharks lay eggs, which sometimes wash ashore in tough protective pouches that are called *mermaids' purses*. Some sharks gestate their young in a yolk sac attached to an umbilical cord and give live birth just like mammals. Very recently born sharks who have developed this way can be identified by their belly button–like umbilical scars. Some sharks reproduce using a strange mix of these methods; the unborn shark grows within an egg, but the egg is inside the mother the whole time. These methods are all points on a continuum, and until 2020, when scientists discovered an entirely new one, there were already ten known distinct reproductive strategies. *Sustained single oviparity* blends elements from *single oviparity* (having one or two eggs at a time in the uterus but not for very long) and *multiple oviparity* (having lots of eggs in the uterus at a time for longer). With sustained single oviparity, there's one egg at a time but it stays in the mother for a longer period of time, which means fewer babies are born but each one has a higher chance of survival because it's more developed.

Sharks also engage in something called *multiple paternity*, in which a female will mate with multiple males during a breeding season, become pregnant by several of them, and give birth to a litter of half-siblings (born of the same mother but with different fathers). Have you heard of *intrauterine cannibalism*, in which some embryonic sharks eat their siblings in the womb? Some scientists believe that this is a response to multiple paternity, aiming to make sure only the strongest genes survive to birth.

Another interesting thing about where baby sharks come from is that several species have been known to exhibit *parthenogenesis*, in which a mother does not mate with a male but is able to become pregnant anyway. She gives birth to a litter of pups that are genetically identical to herself instead of a mix of the DNA of a mom and a dad. This behavior has been discovered in aquariums where female sharks kept in a tank without any males still give birth. It's also been documented in wild populations of some endangered species. It can be an effective way to keep your population's numbers up even if you can't find a suitable mate. Sexual reproduction—mixing the DNA of the female and the male—helps increase the genetic diversity of the species, an advantage that asexual reproduction does not have. Asexual reproduction also results in some genetic copying errors. Scientists have also documented sperm storage in captive female sharks. After mating, a female may decide that now isn't a great time to become pregnant and will simply store the sperm inside her own body until pregnancy sounds like a good option. The record for this type of delayed pregnancy is just over four years.

You have made it through the introductory materials! I'm proud of you. Now that you know sharks are a diverse, widespread, and ancient group of animals with some amazing adaptations for living in the ocean, you'll be well equipped to understand why you don't need to be afraid of sharks, why we're better off with sharks than without them, what's behind some of the threats they face, and what can be done to help them.

2 >> Sharks Are Not a Threat to Humans

Interesting fact: a shark will only attack you if you're wet.

—British comedian Sean Lock

Obviously, this book argues that humans are better off with healthy shark populations than we are without sharks in our waters, and that the benefits of having sharks around outweigh the costs. But implicit in this argument is that there *are* costs, in the form of human injuries or deaths caused by sharks.* Any such injury or death is a tragedy, and I don't want to minimize the suffering endured by shark bite victims or their families. However, it is important to keep in mind the relative risk of such a tragedy when considering policy solutions. As you'll see in this chapter, humans have an incredibly low risk of being bitten by a shark.

Shark bites are one of the topics I get asked about the most during the question and answer sessions of my public talks. It's also the aspect of sharks that I enjoy discussing the least in media interviews. The way many of these media requests are presented sets up the scientist to be the bad guy; my first ever TV interview was on a Charleston, South Carolina, news station after an adorable toddler was (very mildly) bitten by a shark. Saying "Well, actually, shark bites are really rare and sharks are important and in trouble," seems callous when juxtaposed with a cute kid bleeding and crying. (For the record, I suspect that the kid in

*There are also potentially costs in terms of shark depredation, or sharks eating fish caught by fishers, but this is still an emerging issue of concern in need of further research.

question stepped on an oyster or something. I still don't see how a shark bite could have resulted in an injury just to the bottom of the child's foot, but I chose not to propose this in a 30-second spot when I had a clear script to follow.)

In response to the second most common question I'm asked (the most common question is "What's your favorite shark?"), no, I've never been bitten by a shark. I've been bitten *at* a lot. You may have heard that sharks won't bother you unless you bother them first. Well, we scientists often bother them—although we do so for an important purpose, following strict rules of ethical animal research. The first day my father ever came on a research vessel with me (after 15 years of me saying "No, Dad, sharks aren't dangerous; I wouldn't be safer being a lawyer like you") we had to cut the trip short to take me to the ER. This wasn't because of anything a shark did, but because I got careless around our gear and got a large fishing hook stuck clean through my hand. The hook was so big that the hospital didn't have the tools to remove it; a nurse had to run home and borrow a tool from her husband, who ran a landscaping business. I still have the remains of the hook in a medical waste jar on my desk, and a photo of me with the hook through my hand is now part of a government safety training video for fishers. But no, I've never been bitten.

But back to why I don't like discussing shark bites: I frankly don't find

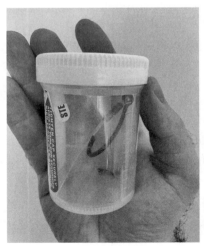

This is what's left of the shark fishing hook that went through my hand. *Courtesy of the author*

the subject particularly interesting. There's an incredibly diverse group of animals older than the rings of Saturn, capable of countless amazing behaviors, critically important to the health of an ecosystem that billions of humans depend on for food, and in serious conservation trouble. Yet even given all that, the topic people want to ask me about is how some sharks bite people occasionally? Perhaps I can interest you instead in a fun, wholesome shark story? The ninja lanternshark, a member of the glow-in-the-dark lanternshark family, was discovered by a colleague named Vicky Vasquez, but was named by her cousin. When Vicky asked for advice at a family gathering about what she should name the new shark, the perceptive youngster correctly pointed out that ninjas are awesome.

What? You still want to talk about shark bites? Fine, let's talk about shark bites.

Shark Bites Are Extremely Unlikely

So how many humans are bitten by sharks in a typical year? It varies a little, but 70 to 100 bites worldwide, of which about 10 are fatal, is a pretty common number, according to the International Shark Attack File (ISAF) at the Florida Museum of Natural History. (In both 2018 and 2019, the total number of shark bites dipped below the historic average, probably due to changes in blacktip shark migrations off the coast of South Florida.) That's 70 to 100(ish) humans bitten out of *BILLIONS* of humans who go in the ocean every year. Most of those are extremely minor bites that require only a band-aid, not surgery or even stitches.

As the satirical news site *The Onion* put it in a 2018 article,

A report released Tuesday by the Woods Hole Oceanographic Institution revealed that only ten individuals will fall victim to fatal shark attacks in 2018; however, you will be one of the victims. Taken as a whole, the number of people killed by sharks has remained fairly low, which is great news for surfers, sailors, divers, and ocean swimmers, but not for you, one of the unlucky ten who will be implacably hunted by the most perfect predator nature has ever devised, dragged beneath the waves so quickly you will be unable to draw a last breath, and torn limb from limb with a savagery that will terrify dolphins ten miles away.

If you've been in the ocean, there was probably a shark near you, and the sharks that were near you almost certainly didn't bother you. That's not because they didn't know you were there—recall from the last chapter how amazing shark senses are. If sharks were the malicious, mindless hunters that so many people wrongly believe them to be, these hundreds of millions of peaceful interactions wouldn't occur.

For a variety of reasons unrelated to sharks, Americans tend to keep much more thorough and easily accessible statistical records on what kills us than many other nations, so I'll mostly use data on the relative risk of death for Americans here. I've taken that data from the ISAF reports, which bounce around somewhat between years and timescales. In a typical year, these reports reveal, one American is killed by a shark. How does that compare with other sources of mortality?

Angry people on my Facebook page regularly assure me that it's not fair to compare some things to the risk of shark bites. After all, almost everyone travels in a car several days a week, so comparing one shark bite fatality a year to the mortality rate of automobile crashes, which kill tens of thousands of Americans annually, is perhaps not a fair comparison. Furthermore, everyone has a heart, and lots of us eat poorly and don't exercise enough, so perhaps it's also not fair to point out that 650,000 Americans die from heart disease every year. Almost everyone goes outside, so approximately 50 deaths a year from lightning strikes is perhaps not a fair comparison with one shark bite fatality a year.

What about deaths specific to the beach or the ocean? In the United States alone, 132 people died at the beach in the year 2000, including 12 who drowned while in areas patrolled by lifeguards. No Americans were killed by sharks that year. From 2004 to 2013, 361 Americans were killed by rip currents, while just 8 were killed by sharks. From the years 1990 to 2006, 11 Americans were killed by sharks, while 16 died by falling into holes on the beach. From 1998 to 2013, just over 3,900

Americans were killed in boating accidents, compared to 13 killed by shark bites. Going to the beach can be dangerous, but not because of the sharks.

There are also plenty of absurd-but-true relative risk of death statistics that, believe it or not, usually result in a chuckle from my audiences. In each of the last few years, more people have died falling off cliffs while taking selfies than have been killed by sharks. Vending machines kill a lot more people than sharks do, as do flowerpots falling from balconies and striking people on the street. Toasters and lawn mowers are basically deathtraps in comparison to sharks. In 1987, New York City alone reported 1,587 cases of humans being bitten by other humans; that year, there were just 13 cases of sharks biting Americans nationwide.

Perhaps my favorite of these relative risk statistics is one that I calculated myself. (Clearly, having a favorite obscure statistic about causes of human death is a sign of a normal and well-adjusted person.) From the years 1580 to 2013, there have been 153 cases in the whole world where a human was undeniably killed by a shark of a confirmed species. The number of fatalities rises to 493 if you count people very likely, but not definitely, killed by a shark of unknown species. In comparison, on the run of the television action drama *24*, we watch Jack Bauer, Kiefer Sutherland's counterterrorist agent character, kill hundreds of people. By my calculations when I initially explored this topic, Jack was responsible for 273 deaths, a figure that has since risen to 309. That's not counting references to his many impressive off-screen exploits; that's 309 people we actually see him kill. In fact, when I wrote a blog post about this entitled "24 Species of Sharks That Have Killed Fewer People than Jack Bauer on *24*," I was thrown for a bit of a loop because, as I discovered, there *aren't* 24 species of sharks that have ever killed a human. It turns out that only 10 species of sharks out of more than 500 have ever killed a human. Five of those species have only ever killed one person.

In the few cases where sharks do bite humans, why do they do it? I don't know, and neither does anyone else, but a leading hypothesis is mistaken identity. This hypothesis claims that the shark doesn't know what we are and so wrongly believes that we're food. The evidence for this is mixed, but the silhouette of a surfer wearing a black neoprene wetsuit looks an awful lot like the silhouette of a seal, natural prey of the great white shark, so it's certainly possible they might initially think that a surfer is a prey item. Additionally, the overwhelming majority of shark bites are known as "hit and run," which means the shark leaves immediately after a single bite—something they hardly ever do when they're actually hunting prey with the goal of consuming it.

Of course, even the few sharks that do injure or kill humans aren't splashing into our homes or workplaces. As intimated in this chapter's opening quote, shark bites almost entirely take place in the water, a habitat that most humans choose to enter for recreational purposes. According to a January 2020 article in the *Palm Beach Post*, there's been an uptick in cookiecutter sharks biting humans. These deepwater animals engage in *micropredation*, taking a chunk out of their prey with their round, cookie-cutter-like jaws instead of eating the entire animal. While this sounds horrific, every one of these victims was swimming across a deepwater trench at night while being lit from above by boats, making them particularly vulnerable and visible. I want to be clear here that I'm not blaming the victims, merely pointing out that when we choose to enter the ocean, we're no longer at the top of the food chain. Anyone venturing into the ocean should be (and usually is) aware of the (very small) risk.*

If Shark Bites Are So Unlikely, Why Are So Many People Terrified of Sharks?

Ignorance is the parent of fear.

—Herman Melville, *Moby Dick*

Shark bites are, statistically, so unlikely that in all functional reality you will never experience one. Chapman University conducts an annual

---

*Legendary shark researcher Dr. Eugenie Clark reported that she was injured by a shark exactly once in her long and storied career: when she slammed on the brakes of her car, a set of preserved shark jaws she had in the backseat launched forward, hitting her arm.

Survey of American Fears in which they ask a random sample of Americans about things they're afraid of. In 2017, sharks were the #41 fear of Americans, with more than 25% of respondents reporting that they are afraid of them. That's tens of millions of people who are afraid of an animal that kills fewer people than being careless while taking selfies. So why are so many people so afraid of sharks?

As reported in a June 27, 2019, *National Geographic* article about the psychology of fear, people are afraid of sharks for a fairly simple reason: because sharks are large wild animals that can hurt or kill you. It makes sense to be afraid of potentially dangerous animals, despite the very small risk. The fact that they usually *don't* hurt people doesn't mean that they *can't* or *won't* hurt you. Humans are hardwired to try and avoid being killed by wild animals, which also explains our fear of things like snakes, which are also extremely unlikely to harm you. In general, humans are really bad at conceptualizing relative risk, something that plagues not only the discourse surrounding sharks but also lots of political issues, including gun control, immigration, and the global war on terrorism.

Since that's a relatively unsatisfying explanation, I'll go into a little more detail about a few factors that exacerbate peoples' fear of sharks: inflammatory, inaccurate popular press coverage in general, the movie *Jaws*, and the dumpster fire of nonsense that is Shark Week.

Inflammatory Media Coverage of Sharks

Sharks are a frequent subject of popular press coverage, and are rarely covered in a positive light. A 2012 *Conservation Biology* article looked at hundreds of examples of sharks being written about in major US or Australian newspapers. The authors found that the most common topic of these articles, by far, was sharks biting humans. More than half of all articles about sharks in major papers from 2000 to 2010 were about a shark bite; only 11% even mentioned shark conservation. The article pointed out that this focus on shark violence is likely to be a problem and suggested that experts make an active effort to speak with the popular press about shark research and conservation topics instead of shark bites.

Eagle-eyed readers may have noticed that I've been using the phrase "shark bites" and not the term "shark attack," which you may be more familiar with. When you hear the phrase "shark attack," you picture the shark from *Jaws*, a malicious creature stalking the coast and killing intentionally simply because it's evil. As we've seen, that's just not what happens; the phrase "shark attack" is therefore misleading and inflammatory. A 2013 paper by Robert Hueter and Christopher Neff instead suggested a new typology of shark–human interaction terms, including "shark sighting," "shark encounter," "shark bite," and "fatal shark bite"; I use their terminology here.

Due to the "if it bleeds, it leads" principle of some unscrupulous strains of journalism, whenever any shark bites anyone anywhere in the world, it's headline news everywhere. This creates the false impression that these events are much more common than they really are, especially when very minor bites get inflammatory coverage. The same 2013 Neff and Hueter paper I mentioned above performed a content analysis of how shark "attacks" were covered in the Australian press, and found a startling statistic: in 38% of reported "shark attacks," THE SHARK DID NOT EVEN TOUCH THE HUMAN. It simply swam near them in a way that the person found threatening or scary.

Sometimes inflammatory media coverage is pretty easy to identify: "Shark Research Makes Us No Safer," "Blood in the Water but Experts Are Still at Sea," "Conservation Policies Value Sharks over Human Lives," "Has Our Admiration for Sharks Gone Too Far?", "Great White Shark: Endangered or Just a Danger to Humans?" These are all headlines from one columnist at one newspaper (Fred Pawle at the *Australian*) that date to the last couple of years. But even the regular language used by journalists who aren't conspiracy theorists can be inflammatory and fear-mongering. Referring to the ocean, which is a shark's home, as "shark-infested waters" suggests that there's something wrong or bad about sharks being there. Referring to wild animals accidentally injuring people as "bloodthirsty" or "monsters" is incorrect and perpetuates public fear and misunderstanding. Similarly, a shark swimming normally and minding its own business is neither "lurking" nor "stalking" humans.

Sometimes popular press coverage is inflammatory even when it's not talking about sharks that bite people. One particularly egregious example of this happened in January 2015, when some Australian fishers caught a frilled shark in their nets. This long and skinny deep-sea dweller has small but sharp teeth and a snake- or eel-like body. It can grow up to six feet long. Headlines about this incident included words like "Horrific" (NPR), "Terrifying" (the *Independent*), and "Like a Horror Movie" (Fox News 2). CNN asked, "What brought this deep-sea monster to the surface?" (It was probably the giant net that it got caught in.)*

Sometimes this media coverage takes the form of misidentifying a species in a way that inspires public fear. Recall that, in addition to shark bites being vanishingly rare, most shark species have never killed a human. One outrageous example of inflammatory coverage appeared in a 2014 *Daily Mail* article, which asked "Is this a great white off the coast of Cornwall?" Even a cursory glance at the image presented showed that it was clearly not a great white—a sometimes-dangerous and fear-inspiring species—but rather a harmless, plankton-eating basking shark. In an article I wrote for *New Scientist* analyzing this particular case, I pointed out a series of major flaws in this *Daily Mail* article. For one thing, the author, Harriet Arkell, didn't interview a single qualified credentialed expert. She *did*, however, interview a fisher who wrongly claimed that the only large fish in UK waters were great whites. Making a common but nonetheless grievous error, Arkell also interviewed a self-described "shark aficionado" (read: someone who thinks sharks are neat but doesn't have any relevant credentials or expertise). As I wrote at the time, "Why quote a shark aficionado, a non-expert who thinks sharks are cool, for a story like this? Can you imagine if journalists did this for other types of story? The White House announced intentions to bomb Islamic State targets in Syria, but counterterrorism aficionado Steve said that he's pretty sure the organization is actually hiding in Peru. Markets cheered the move to reduce interest rates, but finance aficionado John said that everyone should just buy gold and bury it in their backyards. It would never happen, because it's ridiculous."

---

*To its credit, Fox News once issued a correction about a shark article when I contacted them and pointed out a factual inaccuracy.

Similarly, a much-hyped 2013 photo allegedly showed a SHARK IN THE WATER NEAR CHILDREN! Looking at this photo, though, it clearly reveals that the animal in question is not a shark, but a dolphin—which means that an entire week of fear-inducing news was about literally nothing at all. It's perhaps worth noting here that several of the self-described shark experts who claimed this was a shark were non-scientists who regularly appear on Shark Week programming.

Sometimes, this fear-inducing media coverage could just as easily be dubbed "Fish Seen in Water," as in the case of a November 2017 Facebook post by CBS Miami with the headline "Spine-Tingling Swim: Tiger Shark Swims Extremely Close to Miami Beach." The shark in question didn't bother anyone, it was just swimming in its natural habitat. Similarly, a February 2018 article in the *Charlotte Observer* had the headline "A Dangerous Mako Shark Is Haunting NC's Outer Banks and Won't Leave." This shark didn't so much as smile at anyone. It was swimming through its home, but that particular newspaper headline is calculated to frighten. And it "won't leave?" Where do you want it to

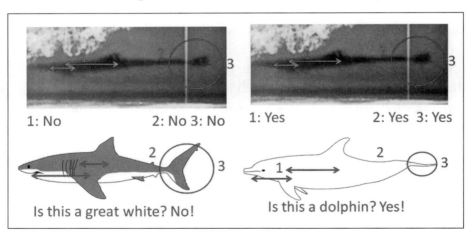

This detail of a 2013 photo, which allegedly depicted a shark swimming near children in the Manhattan Beach surf, was taken by June Emerson, the children's mother. Below, I've used arrows and circles to perform a morphometric analysis by comparing images of sharks and dolphins to the photo. Viewed with this context, we can clearly see that the photo depicts a dolphin, not a shark. *Illustrations courtesy of Kurzon, Wikimedia Commons*

go? It lives in the ocean! Sometimes these articles mention that a shark is "near a beach," which is another way of saying "in the water, which is its home." And while we're talking about this, I'd like to inform Shark Week shows like *Shallow Water Invasion* that this behavior isn't new. Sharks always feed near shore; what's new is everyone has a camera with them all the time, whether it is an iPhone or a drone or a GoPro.

In recent years, particularly flagrant examples of inflammatory media coverage have featured drone footage that shows a shark swimming near humans without bothering them at all. The big story is apparently that the people had a lovely day at the beach without even knowing a shark was near them—isn't that TERRIFYING?

This kind of sensationalist and inaccurate media coverage is a frequent source of frustration for shark scientists, educators, and conservationists.

Jaws

The 1975 Steven Spielberg film *Jaws* totally changed how the world thinks about sharks. Sharks were discussed relatively rarely before it came out, but that summer, people all over the world became terrified to go in the water. My mother reported that she was afraid to go swimming in a *lake* after first seeing *Jaws*. It's hard to think of another fictional film that so thoroughly shaped how the public thinks about a real-world issue, other than maybe *Jurassic Park*. (Picture a dinosaur. You're probably thinking of how they look in *Jurassic Park*, even though we know the majority didn't look like that at all.)

Jaws has been so enormously influential in shaping public understanding (and misunderstanding) of sharks that we refer to the "*Jaws* Effect" in public policy literature. Basically, when there's a high-profile shark bite, local elected officials want to be seen as "doing something," even if that something doesn't really help humans and can even harm threatened species of marine life. Nobody wants to be seen as trivializing the threat of sharks like the Mayor Larry Vaughn character from *Jaws*. On the other hand, Richard Dreyfuss's character, Matt Hooper, the first-ever marine biologist hero in a major motion picture, is cited by many of my colleagues as the inspiration for their own career choice.

A fascinating end to the *Jaws* saga is that Peter Benchley, the author of the novel that the film is based on, was completely horrified by how his story made people want to hurt sharks. He spoke in favor of shark conservation in a series of lectures, funded conservation-friendly documentary films, wrote an ocean conservation book called *Shark Trouble*, and donated much of his profits from the *Jaws* franchise to shark conservation. His wife, Wendy, was on the board of the now-defunct shark conservation nonprofit Shark Savers for many years after his death. And that awesome glow-in-the-dark shark that I mentioned at the start of this chapter, the ninja lanternshark? Its scientific name is *benchleyi*, in recognition of Benchley's shark conservation and outreach efforts.

As for other shark-focused "creature features," like *Mega Shark Versus Giant Octopus* and *Sharknado*, I personally love them and don't think they contribute much to public fear and misunderstanding, in part because they don't take themselves seriously; they're in on the joke. But I should mention that I thank *Sharknado 2: The Second One* in my PhD dissertation. The film's production company, The Asylum, funded some of my research.

Shark Week Nonsense

> Live every week like it's Shark Week!
>
> —Tracy Jordan (Tracy Morgan), *30 Rock*

Shark Week, the weeklong annual series of documentaries on the Discovery Channel, inspires the single largest temporary spike in Americans paying attention to *any* ocean science or conservation topic. It therefore has the potential to be an amazing force for good in terms of promoting science, conservation, and public understanding of sharks. Though they are making improvements, the Discovery Channel has, in general, utterly shirked this responsibility, instead fostering nonsense and fear over facts. Shark Week has long been my own proverbial white whale, and I've been described as "Shark Week's biggest critic" by the *Tampa Bay Times* and the *Miami New Times*. I've even traced several pieces of common misinformation about sharks directly to Shark Week.

Shark Week programs have a long history of presenting stories of so-called shark attacks with melodrama so overwrought that it borders on disrespectful of those bitten. Although shark bite incidents are incredibly rare, the same incident is often recycled on multiple shows throughout the week. Some actual titles of Shark Week documentaries include *Teeth of Death* (1993), *Tiger Shark Attack: Beyond Fear* (2004), *Sharks: Are They Hunting Us?* (2006), and *Ocean of Fear* (2007). And no discussion of Shark Week nonsense is complete without mentioning the horrifically inaccurate *Great White Serial Killer* series, which opens with, I swear to cod, narration saying, "Two shark attacks at the same beach, years apart: Coincidence, or has a shark turned into a serial killer?" (No evidence is ever presented that the same shark is responsible for multiple bites. When I called him out on this, Shark Week producer Jeff Kerr said that they never claim it's just one shark killing multiple people. Which is preposterous; the series isn't called *A Series of Coincidental Bites by Several Unrelated Sharks*.)

Shark Week producers also have a troubling history of lying to scientists to get them to appear on their programs. All too often, they ask a scientist who studies a particular shark species to guest star on a show about that species, but then air content that only peripherally touches on real sharks. Instead, programs focus on fantastical versions of the species that boast preposterous shark superpowers, including the ability to hide from sonar, grow to monstrously enormous sizes, and live far beyond normal shark lifespans. They also misleadingly edit these shows to make it appear that the scientist featured is responding to questions they were never asked. The worst example of this, in my opinion, befell Dr. Kristine Stump, who was asked to appear in a documentary about hammerhead sharks. Dr. Stump was shocked when the show that aired, *Monster Hammerhead*, actually proved to be about a mythical hammerhead dubbed "Old Hitler." It's interesting that the nondisclosure agreements that Discovery has scientists sign to participate in their shows are some of the most thorough and expansive I've ever heard about.

Part of the issue with Shark Week is simply that its content preys on the subconscious, particularly when it comes to the music accompanying what you watch. The background music or score actually matters a lot in all kinds of media, and dark, dramatic, scary background music played when sharks are on the screen makes people associate sharks with being frightening. In a 2016 paper entitled "The Effect of Background Music in Shark Documentaries on Viewers' Perceptions of Sharks," a team of scientists explicitly tested this phenomenon. The same clip of sharks swimming was played to different groups of test subjects, sometimes accompanied by typical ominous sharky music and other times paired with happy, playful, upbeat music, the type typically associated with seeing dolphins on screen. People were more likely to be afraid of sharks and less likely to support shark conservation after seeing the version of the clip with the scary music. I've never seen a group of scientists laugh as hard as they did when Dr. Andy Nosal played shark footage accompanied by the happy dolphin music as part of his presentation at the American Elasmobranch Society conference that year.

Shark Week reached its low point with its series of megalodon specials, starting with *Megalodon: The Monster Shark Lives*. This show claimed that *Carcharocles megalodon*—which you may know either as the giant prehistoric shark whose jaw you can pose in for photos at aquariums or as the CGI star, with Jason Statham, of *The Meg*—is not actually extinct. What's more, this show actually claimed that scientists and the government know that megalodon is still alive and are covering it up. This program's popularity resulted in scientists, including myself, getting death threats from furious and terrified viewers. The show was entirely fictional, using actors as well as CGI videos and photoshopped images that are still trotted out by "Megalodon truthers" to this day. Made by the same production company responsible for Animal Planet's horrific *Mermaids: The Body Found*, this two-hour-long special *never actually said that megalodon wasn't real*, resulting in understandable viewer confusion. In 2013, after *Megalodon* aired, Shark Week's social media team essentially bragged about the fact that they had confused most of their viewers by running a poll in the hopes, as far as I could tell, of further perpetrating disinformation 73% of respondents claimed that "The evidence for megalodon can't be ignored," while only 27% agreed with the statement "No, megalodon is not still alive, the scientists are right." Oy.

Incidentally, this particular bit of pseudoscientific nonsense resulted in my first-ever national TV interview, which had to be restarted seven times. Apparently you can't say the F-word on CNN, no matter how angry you are about incredibly irresponsible and inaccurate shark information airing on the Discovery Channel (sorry again to Jake Tapper and the production staff of *The Lead*). Luckily, *Megalodon: The Monster Shark Lives* was widely slammed in the media. Unfortunately, the next year, Discovery aired *Megalodon: The New Evidence*, which presented actual criticism the first show received as "proof" of "how deep the conspiracy goes!" You know, normal educational TV stuff.

I sent the Discovery Channel a copy of my bar bill the night *The New Evidence* aired. Shockingly, they still haven't paid it. Sadly, against all reason, the myth of the megalodon persists. I speak to thousands of schoolchildren every year about sharks, and I can't recall the last time I *wasn't* asked about whether megalodon is still around. Writing about it here, it's clear that I am still frustrated by this show years later, so let's move on.

Adding insult to injury, although more than 50% of shark researchers are women, you'd never know that from watching Shark Week programs, the highest-profile media coverage my field gets. White men with no degree or credentials are regularly presented as experts in multiple shows each year, while you hardly ever see a woman or a person of color. Even local scientists in countries like the Bahamas, Mexico, or South Africa—where much of the research takes place—are ignored. At a certain point, this kind of exclusion is not a coincidence or an accident anymore. In one particularly disturbing example, the narrator and subtitles of the 2017 show *Great White Babies* referred to a white male who didn't yet have a PhD as "Dr. Elliott," and referred to Dr. Toby Daly-Engel, an actual PhD, as "Toby," never by her title. As my friend Cheng Lee, a computer scientist, said in 2014, "It's sad that the Discovery Channel would rather promote the existence of megalodon than of scientists who aren't white men."

Despite its numerous problems, Shark Week has undoubtedly played a role in helping shark researchers. I, as well as many other scientists, do take advantage of the temporary increase in shark awareness to talk about real science and conservation issues. But there's a lot of bad in Discovery's legacy, and all of it is totally unnecessary. Whatever else it may be, Shark Week is a huge missed opportunity to do some real good for public education about shark science and conservation.

What Can You Do about Shark Bites?

Shark expert David Shiffman says that if you're attacked by a shark, the best way to defend yourself is to attack its what?

(A) Self-esteem   (B) Singing ability

(C) Eyes   (D) Life choices

—*Who Wants to Be a Millionaire*, September 18, 2017 (Yes, this really happened. Obviously, this was the $100 easy starting question.)

People often ask me what they can do to minimize the risk of being bitten by a shark even further. I'm hesitant to even give this kind of advice. When I was a kid, I had information about what to do if I should fall in quicksand drilled into me so many times that I thought quicksand was going to be a much more significant part of my life than it has turned out to be. If you must know, in general, relatively few of the small number of shark bites that occur affect typical beachgoers, who tend to hit the beach in the middle of the day and stay close to shore in a large group. Surfers, the group most commonly bitten by sharks, usually venture out in the early morning or late afternoon. They are often by themselves and far from shore when they are bitten.

Sometimes people ask me what they should do to dislodge a shark that is actively biting them. You don't really have to do anything, and often you can't. Most shark victims report never even seeing the shark that bit them because it so quickly realized that they weren't prey and swam away. Some commonly shared advice is to punch the shark in the nose, but have you ever tried punching underwater? It's great exercise for senior citizens but has little utility for anything else. Note that the correct answer to the *Who Wants to Be a Millionaire* question that opened this section—"If you're attacked by a shark, the best way to defend yourself is to attack its what?"—is "eyes," not "nose," if you ever actually find yourself in this situation. But you (very likely) won't.

There's lots of other common bad advice out there on how to further reduce your risk of shark bites. Perhaps my favorite is the idea that if you see dolphins nearby you're safe to go in the water because everyone knows that sharks are afraid of dolphins. This is nonsense. Dolphins eat the same food as many species of sharks, and the two animals are frequently found near each other. Similarly, there is no evidence of a dolphin ever "protecting" a human from a shark biting them. These stories can usually be more honestly rephrased as "I went in the ocean, I saw a shark, I saw a dolphin, and the shark didn't bite me," which sounds like a pretty normal trip to the ocean that would have turned out exactly the same if the dolphin wasn't in the mix. When I wrote about this misconception for *Slate* magazine in 2014, it resulted in the most hate mail I've ever received for any of my public science engagement writings—and I get a lot of hate mail.

There's also the myth that wearing certain colors, such as so-called "yum yum yellow," (i.e., a shade apparently believed by some to be particularly delicious-looking to sharks) makes shark bites more likely. This is unlikely to be true because most sharks are colorblind. I've also seen people advise beachgoers to avoid wearing shiny material in the water because it looks like fish scales. I'm not sure if there's anything to this but, for what it's worth, I don't swim with my gold necklace.

Shark Culls

> Look, Chief, you can't go off half-cocked looking for vengeance against a fish. That shark isn't evil. It's not a murderer. It's just obeying its own instincts.
>
> —Matt Hooper (Richard Dreyfuss), *Jaws*

Sometimes after high-profile shark bites occur, the "*Jaws* Effect" takes hold and local officials call for a shark cull. In other words, they want to try to make the waters near their beaches safer by killing all the sharks there. This is a terrible idea not only from the standpoint of shark conservation but because we've tried it and it doesn't work. A 1994 article entitled "A Review of Shark Control in Hawaii with Recommendations for Future Research" looked at mid-twentieth-century efforts to kill sharks in an attempt to make Hawaiian beaches safer. The authors found that, despite government efforts to eliminate thousands of sharks that ended up costing taxpayers hundreds of thousands of dollars in 1960s and 1970s money, these activities had no impact on the number of people bitten by sharks at beaches.

The main reason for this is simple: many of the larger shark species are highly migratory. This means that, even if you kill all the sharks in your coastal waters today, by next week, different sharks may be there. In other words, you spent lots of taxpayer money and destroyed threatened, ecologically important animals while making no tangible impact on public safety. The recent cull in Western Australia killed hardly any target species (such as great whites), but did slaughter lots of smaller species that aren't a threat to people. Sometimes these efforts—which usually consist not of *Jaws*-style hunting of one shark, but rather of large-scale efforts to put a bunch of hooks in the water—also kill sea turtles and marine mammals. Shark culls don't make people safer, and they do harm threatened species. We should not do this. We should do something else.

Shark Repellents

Hand me down the shark repellent bat spray!

—Batman (Adam West), *Batman* (1966–1968)

There are lots of so-called shark repellents on the market. The most charitable thing I can say about them is that some repellents work sometimes for some shark species under certain conditions. Many never work, and some are so obviously useless that they're essentially fraudulent.

Most repellent products remind me of the season 7 episode of *The Simpsons*, "Much Apu About Nothing." In it, Homer applauds the high-tech Bear Patrol operation established in response to a lone bear spotted wandering the neighborhood. He exclaims with satisfaction, "Not a bear in sight. The Bear Patrol must be working like a charm!"

Lisa then plucks a stone off the ground and replies, "By your logic, I could claim that this rock keeps tigers away . . . you don't see any tigers around here, do you?" Sure enough, no tigers haunt their all-American hometown of Springfield, but clearly this has everything to do with tigers' habitat requirements and absolutely nothing to do with the rock or its supposed tiger-repelling abilities. This incident made such a great point about specious reasoning that it's been written about in sources like *The American Prospect* and even *Harvard Business Review*.

A 2014 article by the Australian shark conservation nonprofit Support Our Sharks pointed out that just putting a wooden kitchen spoon in your dive gear is exactly as effective at preventing shark bites as most commercial shark repellents and costs much less. If you want to sarcastically make this point to your dive buddies, I sell a waterproof shark repellent sticker on my website. It reads "Official science-based shark repellent sticker: place this on your dive gear and it is extremely unlikely that you will be bitten by a shark, which is also true if you don't use this sticker."

The efficacy of some shark repellent devices, like the Radiator, a stripy wetsuit designed to resemble a visually unappealing sea snake, are questionable. Tiger sharks enjoy eating sea snakes, for one thing. Some shark repellents use magnets or electromagnetic fields, taking advantage of sharks' electrosensitive Ampullae of Lorenzini. In controlled clinical trials, some of these magnetic devices do sometimes seem to mildly pester sharks enough that they don't come back for a second bite at the bait. However, in 2016, someone *wearing* one of these shark repellents was bitten by a shark.

Other Aspects of Beach Safety

The beaches are open and people are having a wonderful time!

—Mayor Larry Vaughn (Murray Hamilton), *Jaws*

There are steps that local governments can take to reduce the risk of sharks biting people, or to reduce the risk of a bite resulting in serious injury or death, without killing sharks. In clear waters, aerial patrols by small plane or drone can identify when large sharks are approaching a beach; then lifeguards or other authorities can warn people to get out of the water. I once had a very strange conversation with Google's Sergey Brin about this strategy. He asked me why we don't just make a network of AI-controlled solar-powered drones to patrol every inch of coastline. After taking a few seconds to compose myself, I pointed out that this solution would cost several times the combined marine scientific research budget of the entire planet.

Depending on local terrain, you also don't necessarily need high-tech solutions like drones or planes. In South Africa, a trained network of volunteer "shark spotters" standing on cliffs with binoculars accomplishes this same task. Of course, that only works when you have high ground for people to stand on near the beach or can afford to install a high lifeguard chair for each spotter. Sometimes resort communities enclose small areas of the ocean to serve as designated swimming areas. This is a very different approach from the one taken in South Africa or Australia, where shark nets are deployed in the hopes of entangling and killing large sharks that approach the beach.

One particularly interesting solution to the problem of potentially dangerous sharks is employed in Recife, Brazil. There, researchers catch large sharks near beaches and simply move them out to sea. By regularly doing this catch-and-release work, sharks are kept away from beaches without being killed. Simultaneously, researchers gain access to lots of study specimens. As documented in a 2013 paper by Fabio Hazin, who sadly passed away due to COVID-19 while I was writing this chapter, simply moving sharks away from people reduced the rate of shark bites off Brazilian beaches by 97%. Meanwhile, threatened shark species survived at a rate of 100%. I had the honor of writing a commentary piece about this article for the journal *Animal Conservation* (when it comes to particularly influential papers, academic journal editors often ask someone influential in that field who wasn't involved in the study to write about what makes the research so important). My commentary piece, entitled "Keeping Swimmers Safe Without Killing Sharks Is a Revolution in Shark Control," extensively praised this approach, calling it "the strategy we've been looking for" and "brilliant in its simplicity." The only problem is that, like every other solution to shark-related problems, catch-and-release doesn't work everywhere. It requires a boat with a crew that is regularly monitoring a set of baited hooks, because if you wait too long the sharks you caught will die. This strategy is just too expensive or impractical to work on huge areas of beach, but it certainly seems like it works wonderfully in Recife.

In general, though, the best way to reduce the (already very small) risk of fatal shark bites is public education. However, as a *Saturday Morning Breakfast Cereal* web comic featuring me points out (see Plate 6 in the color insert), public education has its own issues.

Who Cares if Lots of People Are Afraid of Sharks?

I must not fear. Fear is the mind-killer.

—Frank Herbert, *Dune*

Does it matter that people are afraid of sharks? In short, it matters a little. Some scientific and conservation colleagues question the value of some of my Shark Week outreach. To them I'd say that public fear of sharks is certainly not the biggest or only threat to sharks, but it is more significant than you seem to believe. People's fear of sharks results in less support for shark conservation—that's a scientifically documented fact. In participatory democracies, public opinion influences natural re-source management, including but not limited to shark conservation and management policies. When I give public talks and mention that many species of sharks are threatened with extinction, some people yell "GOOD!" People like that vote.

Public fear of sharks sometimes manifests itself through culls and the "*Jaws* Effect," but perhaps more commonly manifests itself through the fact that elected officials simply don't make shark conservation a high priority. A 2010 paper by Dr. Peter Jacques argues that people's non-chalance about the fate of sharks has resulted in the political margin-alization of shark conservation and management efforts. Jacques wrote that, despite some species of sharks being much more threatened with extinction than less-controversial fishes, sharks get much, much less international conservation and management attention than other species. He asserts that such an obvious mismatch between the degree of the problem and the amount of focus on a solution wouldn't be permitted with a group of animals more beloved by the public. Luckily, since this article was first published, there has been a huge increase in policymaking attention paid to shark conservation, as well as public support for saving sharks. So yes, it matters if people want to save the sharks, and it matters that some people don't want us to save sharks because they're afraid of them.

But public fear and misunderstanding is, in the grand scheme of things, a relatively minor threat. There are many non-expert environmental advocates who focus the entirety of their public outreach on convincing people that sharks aren't actually all that dangerous. This is not a data-driven approach and is therefore much less likely to result in substantive policy change. Sometimes these non-experts engage in wildlife harassment, grabbing, poking, prodding, hugging, kissing, chasing, or even riding wild sharks to prove their point. This strange tactic has the apparent goal of saying, "Look how much I can annoy this animal without it wanting to hurt me, therefore we should save the sharks," which does not make sense to me. Other non-experts claim that sharks are not bloodthirsty monsters (true), but are basically just cute adorable puppy dogs who just need hugs and love (uh, no). Wanting to help sharks and trying to help sharks are both great, but are not the same thing as actually helping sharks.

Public misunderstanding and fear are issues that should be addressed, but there are many other issues that are of much, much greater significance to shark conservation that should be given greater attention. Though the occasional tragedy occurs, sharks are simply not a major threat to humans. The risk of shark bites is so small, and the benefits to having healthy shark populations off our coasts are so impactful, that the choice of how to proceed is clear. And now, with my least favorite subject out of the way, we can talk about those more important issues.

3 » The Ecological Significance of Sharks

In nature, there is something called a food chain. It's where the shark eats a little shark, and the little shark eats a littler shark, and so on and so on until you get down to the single-cell shark.

—Michael Scott (Steve Carell), *The Office*

Soon after a beautiful South African sunrise, a Cape fur seal slowly and uneasily hauls its seven-foot-long, 500-pound body along a rocky shoreline. The five-acre island that this seal and tens of thousands of his closest friends live on is named, cleverly enough, Seal Island. Though Cape fur seals often enter the waters off Seal Island in small, coordinated groups, this one makes the fateful choice to go for a swim by himself. Like other seals and sea lions, Cape fur seals are as fast and graceful in the water as they are hilariously awkward on land. This one is setting off on his morning hunt for fish and squid.

Chris Fallows, a shark ecotourism operator, photographer, and naturalist, beautifully describes what happens next in his book, *Great White and Eminent Grey*:

Deep below the seals, a shark follows like a cork poised to pop, biding its time. Upon an unseen cue, the great fish comes to life, knowing the time to strike is now. With two serpentine-like sweeps of its crescent shaped tail, the massive shark hurtles towards the surface at full speed . . . a thousand kilograms of shark breaks the water's surface and then blasts free, sending gallons of water cascading down its glistening back and streamlined features. In a heartbeat, the shark's raw power, graceful athleticism, and hunter's determination are indelibly imprinted upon those fortunate enough to witness the show. This is unquestionably one of nature's greatest spectacles.

If the seal had survived this initial strike, he might have engaged in a complex defensive maneuver, zigzagging through the water to take advantage of his superior agility in the water. Alas, he'll never get the chance. He is now an ex-Cape fur seal.

Dramatic "nature red in tooth and claw" moments like this one are what many people visualize when they think of sharks as predators, and watching this kind of behavior from Chris Fallows's boat was one of the coolest things I've ever done. While I could pretty happily witness this spectacular behavior over and over again, it's important to remember that sharks are an incredibly diverse group of animals with an equally diverse set of prey and feeding behaviors. (As it turns out, this famous South African great white breaching behavior isn't happening much anymore, a mystery I'll describe later in this chapter.)

If not all sharks dramatically hunt seals, what do sharks eat? It varies a lot. For example, despite being the largest fish in the ocean, whale sharks eat microscopic plankton, slurping up clouds of it and filtering the food out of the water, a feeding strategy familiar to fans of the great baleen whale. (That's why they're called *whale sharks*, along with their size. They are definitely sharks, not whales—and no matter what *Sharknado* co-star Tara Reid says, they are assuredly not half-shark half-whale hybrids.)* Notably, whale sharks and other filter feeders are still predators; eating lots of tiny animals is still eating animals.

Bonnethead sharks, a small hammerhead relative, have been found to digest seagrass, and are the only known omnivorous shark species. We knew that they sometimes ingest seagrass while hunting for crabs in seagrass beds, a process which I've compared to accidentally failing to

---

*After appearing in *Sharknado*, Tara Reid was interviewed on a Shark Week show called *Shark After Dark*. She indeed claimed that whale sharks are hybrids between whales and sharks. Yahoo! News used the perfect headline for this interview: "Tara Reid Confuses Us All with Her Explanation of Whale Sharks."

pick all the lettuce off my deli sub; the fact that some plant matter ends up in my stomach sometimes doesn't make me a vegetarian. But a 2018 study by Dr. Samantha Leigh found that the bonnetheads are able to digest the plant matter, something we've long assumed they couldn't do. This paper, which changed our understanding of sharks dramatically, resulted in one of many times I needed to change this book while writing it. Before Leigh's paper was published, scientists were pretty certain that all sharks were exclusively carnivores.

Tiger sharks, on the other hands, eat sea turtles, shell and all. They have famously been found with all kinds of stuff in their stomachs, including license plates. My favorite unexpected things found in a tiger shark's stomach, though, are penguins. Scientists have even found porcupines in tiger shark stomachs. Tiger sharks have also been known to eat songbirds that grow exhausted during a long migration over water. One time, a tiger shark vomited all over my colleague, Dr. Austin Gallagher. When he looked down and saw some unusual stomach contents splattered on his shirt and bathing suit, he collected and analyzed them, then wrote a paper about what he discovered.

Some sharks are specialist predators, which means that they are picky eaters and only devour one (or a very few) types of prey. This makes them well-adapted to a healthy ecosystem that contains lots of that prey, but vulnerable if something happens to that prey. Other sharks are generalist predators; in other words, they'll eat just about anything. When I was preparing my master's defense presentation on sandbar shark (#BestShark) feeding ecology, I noted with amusement that sandbar sharks eat a large variety of foods, including the study organism of almost every other student in my graduate program.

Shark hunting behavior is as diverse as sharks' diets. Hammerhead sharks use their unusual head shape to pin flat prey animals like stingrays to the seafloor. They also use their extra-wide heads as extra-large surface areas for their extra-powerful electrosensory systems. Since the Ampullae of Lorenzini needed for electrosensing are found on the front of a shark's head, a wider head means more ampullae and a stronger system. Thresher sharks use their tails, which can be as long as the rest of their whole body, to stun fish with a whip-like motion before eating them. Although many sharks prefer to hunt alone, whitetip reef sharks hunt in large groups, wedging their slender bodies into crevices in coral reefs to get at fish trying to hide for the night. Nurse sharks use their unusually shaped teeth and powerful jaws to crush the shells of crustaceans. Angel sharks lie perfectly still just under the sand until prey comes close enough for them to strike.

Most frequently, sharks eat small and medium-sized fishes, as well as crustaceans like crabs and shrimp. But some eat birds. Some eat marine mammals. Some eat squids or octopuses (yes, it's octopuses, not octopi). Some eat sea snakes. So what do sharks eat? Just about anything in the ocean, including other sharks!

While not all shark predation is as awe-inspiring as the breaching behavior of South Africa's great whites, sharks' eating habits can still have important effects on the ocean ecosystems they call home. In biological research, a common way to learn what a particular gene, organ, or body part does for an organism is to break it and see what body systems stop working. (If we turn off Gene X then the animal can't see anymore? Gene X must be associated with sight.) In ecological research, there's a similar principle: to learn about the ecological role of a species, see what happens to the ecosystem when the species isn't there anymore. Sometimes this involves predator exclusion experiments: installing fences that prey can move through but predators can't, for instance. Sadly, though, humanity is engaging in a large-scale natural experiment, creating ecosystem changes by overfishing many shark species all over the world. In addition to some of the results I'll describe here, this sometimes also leads to something called *fisheries-induced evolution* in which the very act of heavily fishing a species causes measurable changes in their biology or behavior, including how fast they grow and reproduce.

In this chapter, I'll describe key ecological concepts associated with the ecosystem role of predators, including examples from terrestrial and marine ecology. The natural world is complicated and many of the examples actually include elements of more than one of these concepts, but I will focus on one concept at a time for ease of understanding. You've already learned that sharks are awesome animals and that they're not the threat to you and your family that alarmist media coverage would have you believe. Now it's time to learn some of the reasons why we're better off with healthy shark populations off our coasts than we are without them.

Predation Release: Predators Keep Prey Populations in Check

One of the most important roles of a predator is to keep prey populations under control. The loss of this control is known as *predation release*. Western Pennsylvania, where I'm from, is currently experiencing serious ecological problems associated with deer overpopulation. Deer populations used to be kept in check by predators like mountain lions and wolves before humans got rid of them (because who wants a big scary predator in their backyard?). However, too many deer can also cause safety hazards for people; a friend in high school once called me to ask for a ride from the mechanic because he had been "hit by a deer" on the highway. Deer are also well-known hosts for diseases like Lyme. This can be an especially large problem now that there isn't enough food in what's left of the forest for all the deer, causing them to venture into human inhabited areas.

If any hometown friends or neighbors are reading this book, first of all, Hi! How are yinz? Secondly, yes, I know that humans hunt deer, an activity so popular in western Pennsylvania that kids get the first day of hunting season off of school. However, while well-managed hunting can control population numbers, it has a very different set of ecological effects than natural predation. Predators target the small, the sick, the weak, and the old among prey species. Hunters instead go for the biggest and strongest to get an impressive trophy, reversing nature's wisdom about ensuring the "survival of the fittest."

Just as wolves and mountain lions used to play a population-limiting role with Pennsylvania deer, water-based predators also help to keep prey populations in check in the ocean. Oceanic predator declines can cause prey population explosions. A 61% decline in starfish-eating fish in Fiji, for example, led to a threefold increase in starfish numbers, according to one study, while another study showed that experimental removal of fish predators led to a 19-times increase in the populations of the colorful flamingo tongue snail on a Caribbean reef.

Declines in shark populations have had similar impacts. One example of predation release that caused a population explosion in shark prey comes from the *pelagic*, or open ocean, habitat of the tropical Pacific Ocean. One resident of this ecosystem is the pelagic stingray *Dasyatis violacea*. Though they look very similar to other stingrays, these animals live in the top 100 meters or so of the open ocean, never reaching the seafloor like their coastal cousins. Pelagic stingrays were once relatively rare in the tropical Pacific; almost none were caught in National Marine Fisheries Service scientific surveys conducted between 1951 and 1958. By the 1990s, when overfishing of their shark predators was well underway, pelagic stingray populations had increased between 10 and 100 times. Population declines in open ocean sharks may have also caused a huge increase in the abundance of large squids.

The removal of reef sharks also seems to have resulted in predation release of fishes, including small rays and moray eels. Sherman et al.'s whimsically named 2020 paper, "When Sharks Are Away, Rays Will Play," found that more rays were observed on coral reefs with fewer sharks present. Those rays were much bolder and less likely to hide from the scientific observation equipment. The latter part of this discovery is interesting because it means that looking for evidence of predation release in complex wild ecosystems like coral reefs is even more complicated than anticipated. Were more rays seen because there ARE more rays, or because the same number of rays are less likely to be hiding from sharks? It's probably a little of both, the paper's authors conclude, but it shows the importance of considering more than one type of data. In 2021, marine ecologist Gina Clementi and her team found that moray eels were more common on reefs with fewer sharks. Coral reefs that were closer to human population centers were more likely to have their shark populations overfished, and reefs with fewer sharks seemed to have many more moray eels. The authors conclude that this increased moray population is due to a combination of (1) moray eels not having to compete with sharks for food, which meant there was more food for the eels, and (2) fewer sharks eating moray eels, which meant lessened moray eel mortality.

The release of *mid-level predators* (animals that aren't at the top of the food chain, but still eat other animals) from predation pressure sometimes results in ripple effects throughout the food web, which are known as *trophic cascades* ("trophic" refers to feeding).

Trophic Cascades

Predators often shape nature in ways that at first glance appear counterintuitive.

—John Terborgh and Jim Estes, eds., foreword from *Trophic Cascades: Predators, Prey, and the Changing Dynamics of Nature*

Sometimes the ecological effects resulting from changes in predator populations ripple through the food chain. This ripple effect is called a *trophic cascade.* The classic example of a trophic cascade comes from the Pacific Northwest. When orca whales began to consume more and more sea otters in the kelp forests of the North Pacific, it wasn't surprising that sea otter populations declined. But the plot thickened! One of sea otters' favorite foods is the sea urchin, which they consume by adorably crushing them with rocks on their bellies (see Plate 7 in the color insert). The population declines of sea otters then resulted in sea urchin predation release. The increasing sea urchin population ate more and more of their preferred food, seaweeds called kelp, resulting in kelp declines. All of this was caused by a change at the top of the food web. Even though orca whales and otters don't eat kelp, changes in how orcas interact with otters significantly affected kelp. And that was bad for everything that lived in the kelp forest.

The most famous example of a trophic cascade in a terrestrial ecosystem occurred in Yellowstone National Park as a result of wolf declines. Fewer wolves meant an increase in the wolf prey population, including giant herbivores like elk. More elk meant more grazing, and perhaps most impactfully, grazing in areas where elk were previously afraid to graze, such as riverbanks that restricted their ability to run away from a predator. This led to major disruptions in a unique Yellowstone ecosystem called an aspen forest. The Yellowstone case study also remains one of the best examples of predator restoration: when wolves were eventually restored, they ate more elk, bringing the population back under control and pushing elk back to their normal feeding grounds. As a result, the aspen forest is growing back.

What about sharks, which are sometimes called the "wolves of the sea"? There are two commonly cited examples of shark-driven trophic cascades. Both are considered fairly controversial in the marine biology world, but I'll explain them here because you're likely to come across them in the conservation discourse. The first, documented in a 2007 paper led by Ram Myers, took place near North Carolina's Outer Banks, where seven species of apex predatory sharks have declined significantly since the 1970s. Sandbar sharks experienced the least decline: 87% since 1972. Declines exceeding 99% since 1972 have been documented among several other species. These declines were believed to result in predation release of small sharks and rays, including the cownose ray. The authors claim that this increase in cownose rays was partially responsible for a collapse in populations of bay scallops, once a commercially important fish in the region, resulting in a shark ⟶ cownose ray ⟶ scallop trophic cascade.

To me, the key message of this study was "Sharks are important and bad things can happen when we overfish them, so let's not do that." Others got a different (and unfortunate) message from this study: "Oh my god, cownose ray populations are exploding. We need to kill them all to save our scallop fishery!" This led to the birth of the "Save the Bay, Eat a Ray" movement. There were even fishing tournaments for cownose rays where anglers used explosive-tipped arrows to shoot at the surface-swimming rays, which is hardly sporting in my opinion. It is unlikely that ray populations could survive this kind of pressure for any extended time, given their very low reproductive rates. I'd argue that trying to solve a conservation crisis by causing another conservation crisis is perhaps not ideal.

It turns out that the data showing this trophic cascade has major flaws in its underlying assumptions, and has been thoroughly rebutted. If you look closely at the data, it would suggest that cownose ray populations supposedly started to increase well after scallop populations began to collapse, almost as if something else caused the scallop populations to decline. (Dean Grubbs, who led the rebuttal, pointed out that this explanation only makes sense if you think that cownose rays can go back in time like the Terminator.) Also, cownose ray populations aren't increasing as much as these data seemed to show. What's instead happening is that existing cownose rays are migrating into new waters. Furthermore, shark populations haven't declined as much as these data seemed to show. The Grubbs rebuttal also notes that including more datasets complicates the supposedly clear pattern shown by the Myers paper. Finally, cownose rays don't really eat very many scallops. So although this is a well-known example of a trophic cascade that is often cited by environmentalists as a reason to protect sharks, it's a fundamentally flawed one.

Another possible shark-driven trophic cascade might operate on coral reefs. Coral animals have a symbiotic relationship with tiny photosynthetic organisms called *zooxanthellae*. They live inside the corals and secrete sugars, which the corals eat. Without exposure to sunlight, zooxanthellae cannot photosynthesize, and the corals will starve. Happily, herbivorous fish like parrotfish help to graze fast-growing algae off of the corals, ensuring that sunlight can reach the zooxanthellae. Parrotfish are eaten by larger fish like grouper, which are eaten by (you guessed it) sharks. The decline of shark populations may cause predation release in grouper, which then eat more and more parrotfish. Fewer parrotfish means more algae growing on coral reefs, which means dying corals.

This model seems to be essentially correct, but it's more complicated than that. It turns out that humans aren't just overfishing the sharks, but also the groupers, and in some cases even the parrotfish. Algae also grows on the coral for because of warmer waters or nutrient blooms, not simply because parrotfish populations are declining. Additionally, corals face other threats besides algae overgrowth. And while it's true that reef sharks often occupy a pretty similar step on the food chain as groupers, some larger sharks eat small and medium sized groupers. So do sharks keep coral reefs healthy? They certainly play important roles in maintaining coral reef health under many circumstances, but as you can see, there are many variables to consider.

A 2013 paper claimed to find true evidence of a trophic cascade on coral reefs. Specifically, the authors argued that Pacific coral reefs that had been heavily fished were home to fewer sharks and more medium-sized predators (called *mesopredators*) than protected reefs, which had more sharks and fewer mesopredators like groupers. On fished reefs with more mesopredators, the authors found fewer herbivorous fishes. Is this a case of a trophic cascade, with declines in sharks indirectly leading to declines in herbivores? Not so fast—a 2016 paper claims that the pattern isn't quite so clear. This rebuttal argues that the difference in shark populations between fished and protected reefs isn't as significant as claimed in the 2013 paper. Furthermore, it argues that some of the fish species the 2013 paper authors counted as mesopredators shouldn't have been treated as such because sharks don't eat those species. That rebuttal got a rebuttal, which got another rebuttal—such is often the way of science. As of this writing, there hasn't been any conclusive evidence of trophic cascades driven by the loss of sharks on coral reefs—in fact, an early 2021 paper found pretty strong evidence of the lack of trophic cascades on the Great Barrier Reef—but the search is ongoing.

Other possible shark trophic cascades include a reef shark ⟶ octopus ⟶ rock lobster food chain. Overfishing reef sharks in Australia seems to have led to an explosion in numbers of their octopus prey, which ate all the rock lobsters and damaged one fishery. Yet another possibly shark-driven trophic cascade involves seals. Fewer sharks means more seals, which eat a lot more fish. Trophic cascades are powerful forces in nature, but they're also really hard to detect because food webs are so large and complicated. I'd guess that even though some of the most popular examples of shark-driven trophic cascades may be flawed, it's very likely that some real cascades caused by sharks are out there.

Some conservation activists have taken things too far, incorrectly asserting that, because of trophic cascades, the crash of shark populations could be directly responsible for the extinction of all life on Earth. According to this argument, which got its highest-profile mention in the documentary *Sharkwater*, phytoplankton, the base of the ocean food web, produce about half of all oxygen on Earth. If we lose sharks, the reasoning goes, this will destabilize the whole ocean, kill all the phytoplankton, and result in the loss of half of all oxygen on the planet, killing everything—including us. Let me note again here that this is not correct, but it's an example of using trophic cascade theory for conservation advocacy.

Trophic cascades are, generally speaking, more likely to occur in simpler ecosystems with more straightforward food chains. If you have five species that serve similar ecological roles as top predators, losing one probably won't disrupt the whole system because the other four can still keep mid-level predator populations in check. If you have only one top predator, losing its ecological role is more likely to disrupt the whole system. The examples described above range from hotly debated to thoroughly debunked, and I share them just to illustrate the general principle despite their particular imperfections. Despite their flaws, these high-profile examples are still useful to think about, if only because something like this is probably happening somewhere.

Keystone Species

Species that have a disproportionate impact on their ecosystem relative to the actual quantity of them living in that ecosystem are called *keystone species*. (A keystone is the top central stone that strengthens and stabilizes an arch; removing it would cause the rest of the arch to collapse.)

So what's the difference between a random member of the ecosystem and a keystone species? The Natural Resources Defense Council, a United States–based organization which has used the keystone species concept in lawsuits to protect the natural world, notes that "one of the defining characteristics of a keystone species is that it fills a critical ecological role that no other species can." They add that "It may not be the largest or most plentiful species in an ecological community, but if a keystone species is removed, it sets off a chain of events that turns the structure and biodiversity of its habitat into something very different. Although all of an ecosystem's many components are intricately linked, these living things play a pivotal role in how their ecosystem functions." So as with trophic cascades, if there's some kind of ecological redundancy between similar species in a complex food web, you're less likely to see these effects. The NRDC notes that keystone species aren't always predators; they can be sources of food or habitat, or even pollinators. They also note that the same species may play the role of a keystone species in one ecosystem but not another.

The first and best example of a keystone predator is a true sea monster. This terrifying beast feeds by prying open the bodies of its prey just enough to eject its own stomach into them. Once they are digested *inside their own bodies*, it slurps up their juices and moves on. Most terrifying of all, if you cut this monster in half, both halves can survive. I'm speaking, of course, about the sea star, colloquially called a starfish, which is absolutely terrifying if you are a clam.

The *Pisaster* sea star is an important predator along the rocky shorelines of the Pacific Northwest, devouring all the small invertebrates in its path, including mussels, barnacles, limpets, and chitons. Since *Pisaster* eats all of these other animals, one might suspect that removing this particular sea star from the ecosystem would be good for these invertebrates. When iconic marine ecologist Dr. Robert Paine tested this theory, what he found shocked the ecology world, starting a whole new field within predator-prey ecology. Without *Pisaster* to keep its numbers under control, mussel populations explode, outcompeting all of the other invertebrates in the rocky intertidal. Limpets, chitons, and barnacles were wiped out in just a few months. This result is called *predator-mediated coexistence*—when the existence of a predator that eats one species makes it possible for another species to exist there too.

Examples showing sharks as keystone predators are limited. The best-known study to explicitly test for this found that sharks in the central North Pacific are not keystone predators because other large fish like tuna are capable of taking over their predatory role. However, you'll often see sharks referred to as keystone predators in environmental ad-vocacy and scientific literature, and they probably are in some cases.

Fear Ecology and the Indirect Effects of Predators

Fear will keep the local systems in line.

—Grand Moff Tarkin (Peter Cushing), *Star Wars Episode IV: A New Hope*

Predators can also have a major effect on their ecosystems by influencing the behavior of prey, a phenomenon known as *fear ecology*. This is one of my favorite ecological concepts, and that's not only because I can show a picture of the Death Star on PowerPoint slides when I teach people about it. This concept is also sometimes called *indirect effects*, because the predator isn't eating the prey, but is influencing it nonetheless.

Fear ecology comes into play when prey species alter their foraging or migratory behavior to minimize the risk of encountering a predator. Have you ever avoided taking a shortcut through a dark alley, spending more time and energy to get somewhere because you thought the alter-nate path would be safer? That's basically fear ecology in action.

For example, dolphins in Western Australia prefer to forage in shal-low, productive waters, but research indicates that they avoid this hab-itat and forage elsewhere during times of the year when tiger sharks are present. Further work in the same location has shown that tiger sharks also influence the foraging behavior of marine herbivores like sea turtles and dugongs (a relative of the manatee). Other research sug-gests that harbor seals in Canada avoid deepwater foraging to minimize the chance of being eaten by deep-dwelling sleeper sharks. Tarpon in the Everglades swim faster and more directly through areas with more bull sharks, cruising through rather than exploring, according to a 2012 paper led by my PhD supervisor, Neil Hammerschlag.

And remember those breaching great white sharks in South Africa from the start of this chapter? Many of them have left the area after decades of being a reliable tourism draw (in addition to a major ecological force shaping their environment). A few years ago, great white sharks started washing up on South African beaches with their livers ripped out, but the rest of their bodies intact. With the wastefulness of shark finning for food and traditional medicine firmly in peoples' minds, many assumed that this violence was a case of human cruelty to wildlife. But it turns out that orcas were targeting sharks, feasting on the protein-rich livers and ignoring the rest of the fish. After this happened a few times, most of the sharks left the region. Finally, after years passed when local tourism operators and scientists essentially never saw any sharks at Seal Island, sevengill sharks started showing up to feast on seals. They had likely been scared away from the region by the presence of great whites. With great whites gone, though, the seal buffet became too tempting to ignore.

The study of fear ecology is a relatively new research undertaking compared to traditional food web ecology. It's even harder to detect the indirect effects of predators than trophic cascades. However, some evidence suggests that the fear ecology effects of sharks may be more common, and more ecologically significant, than the impacts of what sharks eat.

Ecosystem-Wide Effects of Predator Loss

Sometimes trophic cascades can influence not only species within an ecosystem but the structure of an ecosystem itself. Indeed, a scientific paper from 2011 noted that "the loss of [large apex predators] may be humankind's most pervasive impact on nature."

For example, those overpopulated deer in Pennsylvania? Their overgrazing has been partially blamed for declines in forest songbirds, who need underbrush to build nests. The sea stars in Fiji I mentioned happen to be a species that eats coral; their predation release lead to a 35% loss of reef-building corals. The flamingo tongue snail in Florida eats soft corals called gorgonians. Their increased population resulted in an eight times higher than natural rate of gorgonian predation. The kelp eaten by the aforementioned exploding legions of sea urchins would otherwise form enormous, three-dimensional habitats called *kelp forests*; instead, the affected areas have given way to patches of bare rock called *urchin barrens*. When parrotfish don't keep the algae in control, coral reefs, home to thousands of species that can't live anywhere else, die.

Although not as famous or picturesque as coral reefs, seagrass beds are an incredibly important habitat for countless species of fish and invertebrates, including many that support commercially valuable fisheries. Seagrass meadows in Bermuda have significantly declined, a problem that has been partially attributed to predation release of the herbivorous green sea turtle due to tiger shark declines. This kind of overgrazing linked to predation release was used to show that the loss of sharks can exacerbate climate change. How? Fewer sharks eating herbivores means more herbivores grazing. These herbivores release carbon that was otherwise stored in the plants they ate.

As for ecosystem-wide impacts of fear ecology, fish are only willing to travel a certain distance from shelter when predators are nearby, which results in the formation of a halo feeding pattern around rocks and coral heads. Scientists are able to find these *fear halos* in Google Earth images of coral reefs. Fear ecology's effects on herbivores create grazing patterns visible from space. Similarly, *fear release* (when predators disappear, the fear ecology interactions they normally cause disappear, too) can affect an entire ecosystem. In short, ecological interactions are complex and unpredictable. It's impossible to fully envision what happens when you unravel a food web by removing apex predators and keystone species, but it's likely to leave us with a system that's much less able to support biodiversity than what we had before.

Not Always the Top of the Food Chain: Sharks as Prey

Civilization ends at the waterline. Beyond that, we all enter the food chain, and not always right at the top.

—Hunter S. Thompson, *Gonzo Papers, Vol. 2: Generation of Swine: Tales of Shame and Degradation in the '80s*

So far, I've primarily been writing about the ecosystem roles sharks play as predators. But please recall that there are more than 500 species of sharks ranging widely in size. The spiny dogfish, for instance, may be a voracious predator, but at 3–4 feet long, there are bigger fish in their environment. Some of the important ecological roles that sharks play are as prey. Sharks are sometimes eaten by large fish like jacks, grouper, or tunas. Sometimes they're eaten by crocodiles. Sometimes they're even eaten by birds—although the most-shared piece of shark news on social media in 2020, a video which claimed to show a shark being carried away by a bird, really showed a mackerel and not a shark. And as previously mentioned, great whites, the biggest, baddest sharks in the sea, are sometimes snacks for orcas.

A deep-sea grouper eating a Genie's dogfish (*Squalus clarkae*), named after pioneering shark researcher Genie Clark. Photo taken from a NOAA Ocean Explorer submersible. *Courtesy of the National Oceanic and Atmospheric Administration*

A shark conservation campaign graphic highlighting the ecological importance of sharks. *Courtesy of Conservation International*

Why Does Shark Ecosystem Importance Matter for Shark Conservation?

Realizing that sharks aren't that dangerous to humans is sometimes an important first step in convincing people of the need for shark conservation. However, simply acknowledging that these animals are not bad is less powerful than understanding that they're actively good, and that bad things happen without them. This argument is by far the most common one used by shark conservation nonprofits to convince the public of the need to protect sharks. During a survey I conducted of the world's English-speaking shark conservation nonprofits, I found that 63 of these organizations use some version of this argument in their advocacy. The second and third most commonly used arguments for shark conservation—"sharks are especially vulnerable to overfishing so we have to be careful" and "many shark species have declined in population and are threatened"—were used by far fewer nonprofits: 12 and 11, respectively. "We should protect sharks because bad things happen when we lose them" is simply a very effective argument.

The Center for Biological Diversity's shark page states that "sharks are a critical part of the ocean ecosystem, playing an important top-down role in structuring the ecosystem by keeping prey populations in check." The World Wildlife Fund asserts that "these majestic top predators are essential to the natural order of marine ecosystems." Project AWARE notes that "a healthy and abundant ocean, and the communities that rely on it, depend on healthy shark populations." The Hong Kong Shark Federation writes that "sharks are apex predators, they sit at the top of the marine food chain and help to regulate the abundance and trophic diversity of the different species beneath them." The Pew Environment Group had a long-running campaign that stressed the fact that healthy reefs need sharks. There's even a beautifully illustrated children's book about this concept called *If Sharks Disappeared*, by Lily Williams.

Predators play a variety of crucial ecological roles, and the diversity of shark diet and behavior means that sharks are often ecologically vital members of their community. A variety of unpredictable and negative ecological effects can occur when sharks are lost. We are better off with healthy shark populations off our coasts than we are without them for many reasons, including some that haven't been fully analyzed yet. There are risks to having sharks off our coasts, but those risks tend to be *much* smaller than people think. As you've seen in this chapter, the benefits of healthy shark populations are huge. I hope by now that you're convinced that we should have sharks around.

4 » What Are the Threats to Sharks and How Threatened Are They?

How Bad Is It?

In the world of shark conservation, one set of policy solutions are often supported by people who wrongly believe that shark populations have declined more than they actually have, that the consequences of those declines will be worse than the evidence suggests they will be.

—From a 2020 paper I coauthored, "Inaccurate and Biased Global Media Coverage Underlies Public Misunderstanding of Shark Conservation Threats and Solutions"

I'm afraid that I have some bad news for you: many species of sharks have suffered rapid and severe population declines in the past few decades, with an increasing number of species at significant risk of extinction. For a group of animals that has survived several mass extinction events in the Earth's history, this is deeply concerning, even without getting into their ecological importance. Sharks are one of the most threatened groups of animals in the world, and we are at significant risk of losing many species in our lifetimes.

There have been reported declines of 90% or more in some populations of certain shark species since the 1970s. Furthermore, researchers estimate that 90% of all large open-ocean fish—not just sharks but also tuna and swordfish—have declined in abundance. Some of these figures have been disputed, largely for technical reasons having to do with the selection of mathematical models. Still, basically everyone agrees that many species of sharks face grave conservation challenges. It's well documented that some fish species have been completely erased from huge parts of their historical native range. Perhaps the most striking example of range loss is the smalltooth sawfish, a shark-like ray. Sawfish used to be common along much of the east coast from New York to Florida and in the Gulf of Mexico, where they roamed as far west as Texas. Now, they only live in a small area in South Florida. A recent paper about sawfish decline is poetically titled "Ghosts of the Coast."

About one-third of all shark species are considered threatened with extinction by the Red List, which is compiled by the IUCN, a well-respected international group of scientific experts on threatened species. This is an alarmingly high number of threatened species, but it doesn't mean that all sharks are endangered and that we need to enact extreme polices like banning all fishing, everywhere, immediately, as some vocal fringe groups of environmental activists claim. Perhaps the only advantage of the shark conservation crisis's severity is that we don't need to lie to try to convince people that the situation is serious. Accurately describing what is really happening is enough motivation to inspire action.

What Are the Threats?

The diversity of shark habitats, behaviors, and geographic distributions means that sharks face many threats. Some issues endanger many shark species and are categorized as major threats. Others are problems for just a few species or particular populations (subsets of a species found in a specific location). These may be categorized as overall minor threats or local-scale threats. Still other issues aren't a big deal yet but may become a larger problem as conditions change. These are thought of as emerging threats and are something to keep an eye on.

The major threat to sharks is *overfishing*: killing too many sharks, either accidentally or on purpose, to supply a variety of markets for different shark products. There are also a number of minor, local, or emerging threats worth mentioning here, but please keep in mind that these issues aren't currently threatening on a broad species level. This doesn't mean we should ignore these difficulties, but we should keep the scale of the threat in context when discussing how to allocate the limited resources of the conservation movement.

A Note on Conservation versus Animal Welfare

Before we move on to specific threats, I want to mention that what I discuss here will focus on *biodiversity conservation*—keeping populations healthy and preventing species from going extinct—as opposed to *animal welfare*—minimizing the suffering and pain of individual animals through a focus on quality of life and health. Both of these concepts are distinct from *animal rights* arguments, which are rarely applied to sharks. They also differ from *compassionate conservation* arguments, which, to my knowledge, haven't been applied to sharks at all. If you spend enough time among non-expert shark enthusiasts, you'll certainly hear calls to simply ban all fishing because it hurts fish, to "just leave sharks alone," or to ban catch-and-release fishing because it's torture. In my opinion, these are arguments based on personal values, not science. I'm not saying that views that aren't based on science are wrong or bad, although arguments based on personal values that wrongly claim to be based on science are a problem. I'm just saying that arguments based on personal values are not connected to what I do and not what this book is about.

Generally speaking, my work, and that of most of my professional colleagues, focuses on population-level and species-level conservation threats and solutions. Because you're likely to encounter arguments based on animal welfare concerns phrased as conservation issues, I'd like to fully explain the difference between the two concepts. If you want an example of how the goals of animal welfare and the goals of biodiversity conservation aren't the same, I once had a fascinating (albeit extremely troubling) conversation with an animal welfare activist who told me that he doesn't care if species go extinct as long as individuals in that species don't experience unnecessary pain while humans drive their numbers to zero. This is not a common view among the animal welfare crowd, for what it's worth, but it's an illustrative one.

You may be wondering, "Wait, isn't it good to minimize or eliminate animal suffering?" My answer is that of course it is, whenever possible. When the goals of animal welfare conflict with the goals of saving a species from extinction, however, the latter must be prioritized. Luckily, these goals aren't always in conflict. Much of the earliest shark conservation activism, which focused on banning shark finning, combined the ethos and goals of the animal welfare crowd—shark finning is cruel, so we should stop doing it—with the ethos and goals of the biodiversity conservation crowd—shark finning leads to unsustainable overexploitation and population declines, so we should stop doing it. Humane Society International, for instance, one of the most influential animal welfare organizations in the world, remains a reliable partner of many shark conservation campaigns that I've followed over the years. But all too often the goals of biodiversity conservation and the goals of animal welfare seem incompatible.

A lot of animal welfare concerns for fish (including but not limited to sharks) boil down to a simple question: Can fish feel pain? If you ask some of the folks I've had to block on Twitter over the years, they'll tell you that of course fish feel pain because all living things feel pain, and anyone who says otherwise is a cruel monster. Once I was called a Nazi war criminal for questioning this, which is perhaps not a great thing to say to a Jewish man who lost family in the Holocaust.

In reality, animal biology is not so simple. Not all living things, for instance, can feel pain. A lot of living things don't even have a nervous system of any kind. Far too many people seem to assume that all life on Earth consists of a broad variety of small humans with different-textured skin in different-shaped bodies, a problem called *anthropomorphizing*, or attributing human-like characteristics, behavior, or cognition to animals. It's the reason that The Dodo, one of the most-shared Facebook science pages, is explicitly called out in the media literacy assignments I teach my students—it's full of anthropomorphizing nonsense.

Although the "simple vertebrates" like fishes do have a nervous system, they don't have all the same kinds of nerve cells that mammals do—or the same number and configuration of them. The take-home message here is that whether fish feel pain is primarily a matter of the presence and density of certain types of nerve cells and is not influenced by how many ALL CAPS rants you send me on Twitter. I don't know if fish can feel pain, and neither does anyone else. The truth is that if fish can't feel pain, that's no excuse to unnecessarily mistreat them, and if fish can feel pain, that doesn't mean that fishing is suddenly no longer vital for food security and livelihoods in the developing world. From my perspective, the pain sensitivity of fish just doesn't change much on the ground.

There are a few cases in the world of shark conservation where I've seen animal welfare concerns that are either irrelevant from a biodiversity conservation perspective or actively in conflict with biodiversity conservation. One of these cases has to do with telemetry tagging, which is the process of attaching GPS trackers to sharks so that scientists can monitor their migration patterns and behaviors. Some models of telemetry tag are installed by drilling through a shark's fin. If not done properly, this can cause some damage to the fin that persists even after the tag eventually detaches. I know of just one case ever where a shark likely died as a result of a tagging experience, so this is not a population-level threat to sharks. The data gathered from telemetry tagging are important for generating science-based conservation and management plans, so not using tags because of animal welfare concerns actively harms the goals of biodiversity conservation. But for whatever reason, photos of telemetry tagging really bring out the internet zealots. Once I had to get campus security involved when someone threatened to break into my home, kidnap my children, and drill holes through their backs. (I don't have children as of this writing, but still, yikes.)

Telemetry tagging is an inherently non-lethal research activity—after all, you can't track the behavior of a dead animal. Some other research methodologies, however, do require sacrificing the animal. There are certain parts of shark anatomy that you can't sample non-lethally, like the liver, brain, or vertebrae. Such methods are typically used only to generate data necessary for the conservation of an entire population or species. A back-of-the-envelope calculation I did (that is, a rough estimate) suggests that commercial overfishing kills more sharks in two days than have been killed by all scientific research throughout history.

No discussion of this topic would be complete without a brief rant from me about shark dragging. A few years ago, some Florida anglers caught a shark, tied it behind their boat, and drove the boat at full speed while the shark was slowly shredded to death by the force of the water. They also filmed the whole thing and put the footage on the internet. This is gruesome stuff done simply because the humans involved were cruel jerks. However, you should consider two things about this incident. First, this case of one shark being killed cruelly received more media coverage than every other shark conservation story in history combined. Second, the species of shark was not an endangered protected species; it was completely legal for these men to kill this shark, just not in the terrible way they chose to do it. These were not nice people and they did an awful thing, but is what they did worse than the rest of the shark conservation crisis combined? If you agree with me that one shark being killed unnecessarily cruelly is not more important than every other threat to sharks, you'll see why I'm sometimes frustrated by animal welfare arguments in the conservation space.

Overfishing

As I mentioned above, the biggest threat to sharks, by far, is overfishing. This includes target (that is, intentionally fishing for sharks) and incidental (for example, fishing for tuna and accidentally catching sharks as bycatch) catch. It encompasses catch to supply the shark fin market and catch to supply the shark meat market, as well as smaller markets for other shark products I'll cover later in this chapter. It's a big and complex issue, and while no person or organization can focus on the whole problem of overfishing, conservation efforts won't be as helpful if everyone focuses on the same small aspect of the big picture. For instance, in non-expert shark conservation circles, there's an overwhelming focus on the shark fin trade and little awareness that the shark meat trade even exists. Broadly speaking, a focus on shark finning is a focus on *how* sharks are killed, while a focus on overfishing is a focus on *how many* sharks are killed. From the perspective of species-level conservation, how many sharks are killed matters more.

So what exactly is overfishing? There are some technical math-y definitions ($F/Fmsy > 1$), and the definition varies slightly between organizations and agencies. In plain language, we're talking about fishers taking so many fish out of the ecosystem that the ones who are left aren't enough to replace the ones we caught. Eventually, this leads to serious population declines, threatened or endangered status, and possibly even extinction. It is, however, considered unlikely that targeted fisheries alone would cause the total extinction of a species, because once a population declines past a certain point, it costs more in boat fuel to go fishing than you get from selling the fish you catch. Sustainable, well-managed fisheries can catch some fish to sell (and for us to eat), but they leave enough fish in the ocean for those fish to reproduce. It is worth mentioning that a species can also be *functionally extinct*, which means that they still exist but at numbers so low that they can no longer fulfill their ecological roles. Several ray species, including sawfishes and wedgefishes, are already assumed to be locally or regionally extinct, which means that members of their species still exist in the wild somewhere in the world, just no longer in that part of their historic range.

"Overfishing" refers to an ongoing action (we are currently taking too many fish out of the system), while "overfished" refers to a current state (there are too few fish left in the system). A given fish stock can be overfished with overfishing still occurring, which is particularly bad. A stock can also be one but not the other: overfished already, but without ongoing overfishing; or not yet overfished, but undergoing overfishing. The goal is for a fish stock to be neither overfished, nor suffering from ongoing overfishing. This is visualized in Fisheries Science World™ on a figure called a Kobe plot, which you can see below; a sustainable fishery would be in the lower right quadrant (no overfishing occurring, not overfished) (you'll often see this represented as green in fisheries technical reports).

As discussed earlier in this book, sharks' life history—that they have relatively few babies, which they bear relatively late in life and relatively infrequently—makes them inherently vulnerable to overfishing. Please understand that this doesn't mean all sharks are endangered. There aren't very many statements that are true for all species of such a diverse group

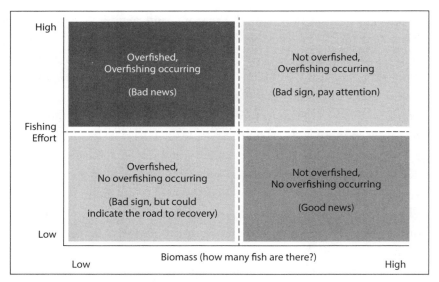

A mockup Kobe plot. A real plot would be covered in data points indicating the outputs of various models and would be accompanied by a long technical report. It would also indicate what percentage of those model outputs fall into each quadrant. *Courtesy of the author*

of animals. It also doesn't mean that sustainable fisheries for sharks are impossible or that we have to ban all shark fishing across the world in order to restore shark populations to a healthy level.

So which sharks are currently overfished with overfishing still occurring—the worst possible scenario? To give you a sense of how this is assessed, in the United States, the National Oceanic and Atmospheric Administration, or NOAA, produces a document every year called the Stock Assessment and Fisheries Evaluation (SAFE) report. As of 2019 (when the most recent report as of this writing was released), here's the status of some of the most economically valuable fished or protected sharks in US Atlantic and Gulf waters (not counting spiny dogfish which are managed separately, and are notably the biggest shark fishery in the USA). You'll see designations of *no/no* (not overfished/overfishing not still occurring), *yes/yes* (overfished/overfishing still occurring), and *no/yes* (not overfished/overfishing now occurring), or *yes/no* (overfished/ overfishing no longer occurring).

Table 4.1 The status of Atlantic and Gulf of Mexico shark stocks, from the NOAA SAFE (Stock Assessment and Fisheries Evaluation) report from 2019

Species	Overfished in US waters?	Overfishing occurring in US waters?
Porbeagle (NW Atlantic)	Yes	No
Blue shark (North Atlantic)	No	No
Shortfin mako (North Atlantic)	Yes	Yes
Sandbar sharks	Yes	No
Blacktip (Gulf of Mexico)	No	No
Blacktip (Atlantic)	?	?
Dusky sharks	Yes	Yes
Scalloped hammerhead	Yes	Yes
Bonnethead (Atlantic)	?	?
Bonnethead (Gulf)	?	?
Atlantic sharpnose (Atlantic)	?	No
Atlantic sharpnose (Gulf)	?	No
Blacknose (Atlantic)	Yes	Yes
Blacknose (Gulf)	?	?
Finetooth	No	No
Smooth dogfish (Atlantic)	No	No
Smooth-hound complex	No	No

There are indeed quite a few shark stocks that are overfished, with overfishing still occurring. You will notice question marks in the table. These mean that we have no idea if this species has been or is being overfished because a formal *stock assessment* (a scientific and technical population analysis) hasn't been conducted. In some cases, the question marks appear because there *was* a stock assessment, but it didn't produce conclusive results. This lack of conclusive data stands out, especially for the United States, a country that (often correctly) bills itself as having some of the most sustainable, well-managed shark fisheries in the world. A report from the nonprofit group Oceana quantified this using 2017 numbers from a past SAFE report and other similar documents. They found that about two-thirds of all US shark stocks have unknown status; 19% are not overfished, overfishing not occurring; 12.5% are overfished with no overfishing, or vice versa; and 6% are overfished with overfishing. That sounds bad.

When you weight this data by the actual amount (in pounds) of sharks caught, however, not just by how many species exist, the picture looks very different. In 2017, 2.4% of sharks caught in US fisheries came from stocks with unknown status, 0.2% came from landings (that is, they were caught by fisheries and brought to shore to sale) with mixed status, and 0.4% came from stocks that are overfished and experiencing overfishing. This means that 96.6% of landings came from stocks that weren't overfished or experiencing overfishing. Also, of the 40 stocks with unknown status, it's already illegal to retain 18 of them if caught. So is there more work to be done to further improve the sustainability of US shark fisheries? Absolutely there is. But it's notably less of a problem that lots of shark stocks haven't had a stock assessment conducted once you recognize that those shark stocks are a fairly negligible part of the overall shark fishery.

The SAFE report includes an estimate of how long it'll take for particular species to recover. These estimates range from 10 years (for scalloped hammerheads) to more than 100 years (for dusky sharks). There aren't too many government documents that plan for over a century into the future, so dusky sharks got that goin' for them, which is nice.

How Many Sharks Are Killed Every Year?

For we are all killers, on land and on sea; Bonapartes and Sharks
included.

—Herman Melville, *Moby-Dick*

Truth be told, we just don't know how many sharks are killed every year
by humans. The best comprehensive estimate out there comes from a
2013 paper led by Boris Worm. Before you go shouting these numbers
from the rooftops, be aware that this paper is essentially a back-of-the-
envelope calculation that makes a lot of assumptions and uses a lot of not
especially reliable data. Many of my colleagues roll their eyes when it's
mentioned. But when not especially reliable data are the only data available, your options are either to use that data or to give up on your project
altogether. I'll therefore report on what the paper authors found, while
stressing that you should take these findings with a large grain of salt.

With those caveats out of the way, here's what they discovered: between 63 million and 273 million sharks a year are killed by humans.
That sounds like a lot, but again keep in mind that the devil's in the
details. Killing ten thousand Natal shysharks is extinction; killing ten
thousand gummy sharks, on the other hand, is about a month's haul in
a well-managed sustainable fishery (gummy sharks are evaluated by the
Australian fisheries management authority as not overfished, with no
overfishing occurring).

Can you name the top ten shark fishing nations in the world, which
together account for about half of the reported shark catch? According
to the 2015 Food and Agriculture Organization of the United Nations
(UNFAO) report, "State of the Global Market for Shark Products," they
are, in descending order, Indonesia, India, Spain, Taiwan, Argentina,
Mexico, the United States, Pakistan, Malaysia, and Japan.

Incidental Catch and Bycatch

Bycatch refers to fishers accidentally catching something just because it
is swimming near what they're really trying to catch. When fishing for

fun with a rod, you catch one fish at a time, and if it's not what you wanted you can just throw it back. This is just not possible with modern industrial-scale fishing gear, which sometimes involves dragging miles-long nets through the water or using tens of thousands of baited hooks at a time. Even a big school of your target catch is going to have some other animals among it.

In some fisheries, there's more bycatch than catch of the actual target species, which is technically called *low selectivity*, or colloquially referred to as "pretty messed up." Sometimes bycatch is also sold by fishers, and may even be desirable to them. Other times, bycatch is of little or no value, or is protected legally, so fishers don't want it; conditions on the ground are always messier than they are in textbooks. You will some-times see references to *unintended catch* or *non-target catch*; it's all ba-sically the same issue, though often with different causes and different solutions.

Bycatch is a significant threat to sharks, although it's not an insur-mountable one. When we fish for species that sharks eat, it's pretty likely that we'll find sharks near those fish. This is such a big problem that the textbook *Marine Conservation Biology* notes, "Our chief concern with fishing as an extinction risk is not directed fishing, but the elimination of many other species that are not the primary targets. Because fishery management is typically geared to the target species, the disappearance of large, conspicuous bycatch species can go unchecked and unnoticed." A UN report from 1994 claimed that over one-third of all sharks killed by fishing were originally caught as bycatch, while a 2000 paper claimed that the number was closer to half of all sharks killed by fisheries. The World Wildlife Fund's bycatch website mentions that, "in terms of numbers, sharks are the most significant bycatch species in the world's major high seas fisheries." And these alarming numbers don't count the many, many sharks that are accidentally caught by anglers and released but still die as a result of capture stress.

Sharks (and rays) are caught as bycatch in fisheries using basically every gear type, and several of the highest-profile shark conservation issues involve bycatch. Mako sharks, dusky sharks, hammerhead sharks, and blue sharks are caught as bycatch in tuna and swordfish longline fisheries, which consist of thousands of baited hooks. In the 1990s, dusky sharks were the second most commonly caught shark in US long-line fisheries. Critically Endangered angel sharks were caught as bycatch in bottom trawl fisheries, which drag a heavy net across the seafloor to catch species like scallops, shrimp, or flounder. Juvenile great whites were caught in California gillnet fisheries, which are large stationary nets that fish swim into. Although bycatch is only one subset of over-fishing, it's still a case of humans killing too many sharks via fishing.

Shark Finning and the Shark Fin Trade

"Shark Fin Soup: The Toxic Delicacy Causing Ecosystem Chaos"

—The somewhat provocative title of a 2019 CNN article

Many passionate but non-expert shark conservation activists have pri-marily heard about one threat to sharks: shark fin soup. Unfortunately, much of what you've heard isn't correct. Indeed, some of the most frus-trating conversations I've had on social media have been with passion-ate but uninformed sharkophiles who are simply incorrect about what shark finning means and does not mean. When my coauthors and I argued in our 2020 paper on inaccurate media coverage of shark con-servation threats that "shark conservation threats and their solutions are frequently presented in an oversimplified, biased, or factually inaccurate manner that would likely contribute to widespread public misunder-standing about these topics," this was what we were worried about.

Shark finning refers exclusively to the act of a fisher catching a shark, cutting the fins off of that shark, and discarding the carcass at sea. If it's still alive, the shark will bleed to death or suffocate (recall that many sharks need to swim to breathe) over the next few hours. This practice is widely perceived as especially inhumane whether or not sharks feel pain. It is also wasteful and is rightfully condemned by a broad coalition of groups, including welfare activists, environmentalists, scientists, and even some sustainability-minded industry groups. Happily, this public condemnation has proved effective. Finning is happening much less than it used to, and has been totally banned in more than 40 countries, including most of the largest shark fishing nations. It's been illegal in the United States since the early 1990s, something you should keep in mind the next time one of your scuba diving buddies asks you to share an online petition calling on the state of Florida to ban shark finning.

A lot of people who I have these frustrating conversations with on social media wrongly use the term "shark finning" to refer to the practice of anyone killing a shark for any reason or using any fishing method to do so. But shark finning, as we've seen, has a narrow and specific definition. If a shark's carcass is brought to land, that shark has unequivocally not been finned. Even if the fins are later removed and sold, and even if the carcass is not used for anything, it doesn't count as shark finning if the carcass has been brought to land. To qualify as a case of shark finning, the carcass must be dumped at sea. This means that, among other things, it is absolutely possible that there are shark fins sold that did not come from shark finning, and it is absolutely possible that there are shark fins sold that came from a sustainable, well-managed fishery. (Repeatedly pointing this out has led to angry accusations that I support shark finning, which is so ridiculous as to not deserve a response beyond rolling my eyes.)

But let's back up a little: Why harvest shark fins? Who wants them and what are they used for? Shark fins are used to make a traditional delicacy associated with celebration in China and Chinese diaspora communities worldwide called shark fin soup. *National Geographic* claims that this dish can be traced back to an emperor from the Song Dynasty who lived over a thousand years ago. Other reports place the dish's origin a few centuries later, but it's still much, much older than the founding of the United States. The emperor and his court consumed this and similar difficult-to-get dishes as a sign of status and mastery over the world around them.

Notably, the fin component in shark fin soup has no flavor or nutritional value. Fibers from the fin called *ceratotrichia* transform when cooked into a noodle-like substance, giving texture to a broth that gets its flavoring elsewhere. In the past, a few dozen members of the imperial court indulging hardly constituted a shark conservation crisis. But following China's economic boom in the 1980s, the emerging middle class wanted to demonstrate how wealthy and powerful they'd become, resulting in a massive increase in demand for this traditional but formerly rarely consumed cultural dish. Today, it's often served at weddings and business dinners and such. Though just about any fin works (and I'm assured that you can't really tell the difference in the final product), because fins from rarer species are harder to get and more valuable, incorporating them into a dish offers people a greater opportunity to show off their wealth and success.

We know that the practice of shark finning is distinct from the shark fin trade, which involves selling shark fins that may or may not have come from shark finning. These two practices are often discussed together, though, so I'll follow suit.

How many sharks are killed by shark finning? We don't know. A commonly cited but not quite accurate figure, 73 million sharks a year, comes from the high end of a 2006 estimate by Dr. Shelley Clarke, who did some of the earliest and most influential work on the conservation issues associated with the shark fin trade. She estimated that the fins of 26 to 73 million sharks a year pass through the fin trade. This is *not* the number of sharks "killed for their fins" because it includes sharks that were killed for meat and also had fins sold. It also includes bycatch species that weren't killed on purpose. This distinction isn't mere pedantry; these are different problems with different solutions. This misuse of Dr. Clarke's data by well-intentioned activists claiming that this was the number of sharks killed for their fins so frustrated Dr. Clarke that she wrote an op-ed about it for the ocean education website SeaWeb. In her piece, she stressed that "selective and slanted use of information devalues and marginalizes researchers who are working hard to impartially present the data." And remember that 2013 back-of-the-envelope calculation mentioned earlier? It estimated that about half of estimated shark mortality in the year 2000 (908,000 metric tons out of 1,638,000 metric tons) was associated with shark finning. Subsequent papers indicate that this figure has declined a lot in the past 20 years as more countries and fisheries management organizations ban shark finning (which again, does not mean banning the sale of shark fins). So it's fair to say

"The fins of up to 73 million sharks are traded every year," or "The fins of tens of millions of sharks are traded every year," but not accurate to say "73 million sharks are killed for their fins every year." And it's super not accurate to say "100 million sharks a year are killed for their fins," or "273 million sharks a year are killed for their fins," as that conflates several different numbers while using some of them in the wrong way.

The shark fin trade is often associated with Asia. Western conservation activists usually think of it as a far-away problem that's other people's fault and which other people must work to solve. This othering often leads to some extremely racist discourse I won't repeat here. Regardless, the idea of a distant problem that "we have nothing to do with" is true of almost no issue in our increasingly globalized world, and it's certainly not true of the shark fin trade. More than 100 countries export shark products, including fins, making this a global issue that demands global solutions—one that will require countries to actually follow through on global commitments. It's true that eight of the top ten shark fin exporting nations, which include China, Thailand, Indonesia, Taiwan, Singapore, Vietnam, the United Arab Emirates, Yemen, and Malaysia, are Asian. But the United States isn't far behind, coming in at #13. Other top 20 countries include Trinidad and Tobago and Brazil.

There is cause for hope. Although shark fins are still quite valuable and commonly traded around the world, consumption has decreased precipitously and imports into Hong Kong are declining. This has been attributed to a few factors, including aggressive environmental and animal rights advocacy, declining taste among younger generations of Chinese people (which may or may not be associated with environmental awareness),* and concerns about the health risks associated with shark fins, which contain higher-than-recommended levels of toxins like mercury. However, at least some of this reported decline has to do with changes in supply chains and international customs codes. This implies that it isn't necessarily that fewer sharks are being killed, but simply that we have altered the way we record how many sharks are killed.

---

*Though the reported spectacular success of a shark fin soup reduction campaign involving Chinese NBA superstar Yao Ming has been challenged in the academic literature.

The Shark Meat Trade

You can be forgiven for believing that the only reason people kill sharks is for their fins. The truth, though, is much more complicated. A global analysis I coauthored of how shark conservation is portrayed in the media found that the shark fin trade was mentioned in 3.5 times as many articles as the shark meat trade. About half of all mentions of the shark meat trade were factually inaccurate assertions that there is no shark meat trade. In reality, demand for shark meat is large and growing, up more than 40% in the past couple of decades according to the UN "State of the Global Market for Shark Products" report I mentioned earlier. Shark is hardly a rare delicacy. The Publix supermarket next to my parents' retirement community sells multiple different types of shark meat at the seafood counter. But it's important to remember that local small-scale shark fisheries around the world contribute significantly to food security and local livelihoods. The desperately poor probably can't just order a kale salad for lunch. If your environmental activism doesn't acknowledge that many people may not have a lot of viable alternative options, it's not especially likely to succeed. And I can't emphasize enough that the much larger shark meat trade is not addressed at all by calls to simply ban shark finning or ban the sale of shark fins, specifically. The significant impacts of these kinds of nuances demonstrate why it's important to understand the complexities of threats to sharks. You can't solve a whole big problem if everyone just addresses the same one small part!

While I don't encourage people to go out of their way eat shark, if the meat comes from a sustainable and well-managed fishery I don't have a problem with it. I've eaten shark meat, and have cooked it for friends (a colleague had sacrificed some finetooth sharks for his research and distributed the meat around the lab so it wouldn't go to waste). It's kind of tough and groupery. Have you ever noticed how about 99% of commonly sold fish can be described as "mild flaky whitefish, not too fishy"? If you prefer that kind of flavor, you probably wouldn't enjoy shark. Like just about anything, though, it can taste good if you prepare it right.

Minor, Local, and Emerging Threats

While the largest threat to sharks overall is overfishing fueled by demand for meat and fins, other parts of sharks are sometimes sold. Some of these relatively minor threats are commonly misrepresented as major when they are absolutely not. For example, a recent amateur online petition protested the supposedly huge threat posed by the sale of dead baby sharks in formaldehyde-filled jars, a bizarre tourist curio. Selling baby sharks in jars, while creepy and weird, is not a major conservation concern, though that petition, proclaiming "Species will never recover if we kill the babies too," garnered over 7,000 signatures. (In reality, these babies were bycatch in a sustainable fishery. Neither the bycatch nor their sale is a serious conservation issue.)

While we're talking about weird stuff for sale at tourist shops, I'm sometimes asked about those "free shark tooth necklace with purchase" tchotchkes that you can get at most roadside stores in beach towns. Because a single shark can go through thousands of teeth in its lifetime, shark tooth necklaces are not a shark conservation concern. Nor is the use of black shark teeth in jewelry an ethical issue: the color reveals that these teeth are fossilized and any shark they came from has been dead for tens of thousands of years.

What about shark jaws, which are also sometimes sold to tourists? Whether this practice is a concern depends on the species that the jaw in question comes from. Some of these jaws come from endangered protected species, which is obviously not good. Others are the byproducts of sharks killed as part of sustainable, well-managed fisheries. As for other consumer uses, so far, there aren't a significant number of sharks being killed to provide skin to make clothing and shoes. Shark cartilage and shark liver oil (which is sometimes used to make the compound squalene, needed for various pharmaceutical products, including some COVID-19 vaccines) are typically extracted from sharks that are already killed by fisheries.

Some popular "alternative" medicines are also made from sharks. These products are nonsense. Shark cartilage, for example, is peddled as a cancer treatment by unscrupulous people who falsely claim that sharks can cure the disease because they don't suffer from it. First off, sharks absolutely do get cancer. Of course, even if sharks were immune to cancer, eating shark wouldn't cure your cancer any more than eating LeBron James would make me better at basketball. Anyone making this claim is lying to and exploiting desperate and ill humans. Luckily, this kind of quackery is currently not associated with major increases in demand for shark exploitation.

Recreational Angling

Bring your big boy pants.

—A Florida charter boat fishing website describing the thrill of recreational shark fishing

Recreational fishing, or fishing for fun, rather than as a primary source of income, isn't commonly considered a major threat to sharks. Heck, recreational anglers are often (mostly accurately) described as partners for conservation rather than conservation threats because they want healthy oceans just like environmentalists do. Indeed, the overfishing discussed above only includes commercial fishing. There's no doubt that industrial-scale fishing techniques, not someone going fishing for fun with their grandpa on the weekend, are the reason that many shark species are threatened. Now that many species are already threatened due to overfishing, however, some populations of particular species are low enough that recreational fishing can have a big impact.

If you've seen videos of the massive amounts of fish caught by industrial fishing vessels, comparing this with your own meager catch after a day on the water may lead you to wonder how this is possible. Well, sure, one commercial fishing vessel can catch a lot more fish than one person with a rod and reel. But there are a *lot* more people with rods and reels than there are commercial fishing vessels. Tens of millions of people fish for fun in the United States alone, and an estimated 220 million people worldwide fish recreationally every year. Worldwide, these anglers land about fifty billion fish a year, not counting fish that they release that die as a result of the stress of being caught (this is called *post-release mortality*), according to a 2004 analysis. Another paper published the same year found that more than a third of total catch of some species of conservation concern comes from the recreational sector.

Recreational fishing is a potential conservation concern due to the economic forces driving the activity. Although anglers aren't trying to catch so many fish that they pay for the trip, they *do* pay for the experience of catching a fish. Usually, the rarer the fish, the more anglers are willing to pay for the chance to catch it.

Additional factors make recreational fishing even more concerning. There are many species of fish with populations so low that commercial fishing for them is banned, yet recreational fishing of those species is still allowed. A 2019 paper found that the majority of countries don't manage their recreational fisheries very well, and many don't manage them at all. Of course, failure to manage an environmental threat to any natural resource creates a much greater likelihood of the collapse of that resource.

Some recreational anglers become obsessed with trying to land the largest fish ever caught of a certain species, with the goal of earning a prestigious award from the International Game Fish Association (IGFA). This is known as trophy fishing, and the mentality behind it is as different from most recreational fishing as the difference between hunting for deer in the forest near your home in order to eat their meat versus flying halfway around the world to hunt an endangered mountain goat in pursuit of some "exotic" décor to put over your fireplace. IGFA world records are problematic in some cases because they implicitly require anglers to kill the fish. You aren't eligible for an all-tackle weight record unless you bring the fish back to shore and weigh it at an IGFA certified weigh station.

The larger the fish, the more detrimental its death can be. As noted in a 2014 analysis I led, targeting the largest members of a fish species has a disproportionate impact on that species' population dynamics. This is because in many fishes, a larger mother is able to have more and healthier babies than a smaller mother. For example, a single 61-centimeter-long red snapper can have as much reproductive output as over 200 41-centimeter-long red snappers.

Our analysis also found that IGFA further compounds the issue by offering awards for 85 species of fish considered threatened with extinction by the IUCN Red List, including more than a dozen threatened species of sharks. Our solution to this, by the way, was for IGFA to simply not offer records for threatened species anymore, which would still allow them to operate for 93% of all species in their 2014 record book. I described this proposal at an international conservation science conference as the lowest-hanging conservation fruit in the world, because, I argued, there isn't a single other conservation solution anywhere that can so easily help so many threatened species around the globe for such a negligible cost. As of this writing, a petition started by the Blue Planet Society asking IGFA to follow our recommendations has been signed by over 70,000 people. The signatories include many people who have publicly said things like, "I'm not anti-fishing, I love fishing and have been fishing my whole life, but killing the biggest members of endangered species is no good." Thus far, IGFA's response to our suggestion has been less than warm.

Although what you've read so far provides a general overview of conservation issues associated with recreational fishing, it still begs the question, What about recreational fishing for sharks? In graduate school, when I first started looking at the possible impacts of recreational shark fishing on threatened populations, some of my respected senior colleagues reached out to me to say that I shouldn't waste my time, since everyone knows that commercial overfishing is the problem, not recreational fishing. (Notably, this is not a thing that typically happens to graduate students and was a consequence of my relatively high profile on social media and such.) After some great discussions, we agreed that though there wasn't a whole lot of evidence for this yet, maybe it's enough of a problem that it's worth one graduate student taking a semester to look into it. Spoiler: it is a very good feeling to have someone you respect a great deal tell you after a debate that, in fact, you were right. It turns out that recreational fishing is, for some populations of some species, a much greater conservation threat than was previously realized, and is certainly worthy of considerably more research, management, and advocacy attention than it has received to date.

I've given scores of professional talks at more than 60 scientific conferences all over the world, but I've only had professional colleagues audibly gasp in the middle of one of my presentations once. I wasn't even presenting one of my results, I was just introducing the problem before describing what I did next. I simply pointed out that, according to NOAA's annual "Fisheries of the United States" report for 2013, more large sharks were killed in US waters by recreational anglers (about 4.5 million pounds of shark) than by commercial fishers (about 3.2 million pounds of shark). Although these figures came directly from a well-known public report that many of my colleagues regularly cite, I was apparently the first person in my professional circle to compare the two relevant tables side by side (the report is massive and contains hundreds of tables, so everyone just sort of reads the part that is relevant to their own work). And further scrutiny has shown that more large sharks in the United States have been killed by recreational anglers than commercial fishers every year since that report was published. This pattern alone shows that it's probably worth paying a little more attention to recreational angling. So we did.*

The presentation that audibly shocked many in the audience was about some of the results from a study my colleagues and I conducted on charter fishing captains in Florida by analyzing their websites and advertisements. We also sent these captains a voluntary survey. We found that there are a *lot* of charter fishing operations in Florida that target sharks, and that sharks are a significant part of these captains' business. We also found that most charter captains in Florida require that their clients always or almost always practice catch and release when shark fishing, which is great news for some threatened species. (Incidentally, the captains who said they "almost always" practice catch and release asserted that the only times they don't release the sharks they catch are

---

*The "large shark" part here is important. About 18 million pounds of dogfish were killed by commercial fishers, but rightly or wrongly, conservation attention often focuses on the bigger species.

when they are going for an IGFA record.) Finally, we discovered that charter captains value having healthy shark populations, both for economic reasons (for instance, "quick and reliable fishing action makes for a better charter business") and ecological reasons ("sharks keep the ocean in balance, and we need to respect the marine ecosystem"). In terms of catch-and-release fishing, captains said things like "There's no need to kill something you're not going to eat," "Why kill them? I respect the sharks," and my favorite statement quoted in my whole PhD dissertation, "Times have changed. The *Jaws* craze is over, and with greater public consciousness toward conservation, catch and release has become the norm." (It was a fisherman in rural Florida who said this, not a Greenpeace activist.)

My greatest professional regret to date relates to this charter fishing study. In it, I mentioned that one of my surveyed charter captains reported regularly catching leopard sharks in the Florida Keys. This species' skin is covered in vibrant spots, making it quite an attractive catch, but it does not live in the Atlantic Ocean. Intriguingly, there's nothing that looks remotely like a leopard shark in Florida waters. This captain even invited me to come down and fish with him to see what he was talking about. I wish I'd taken him up on the offer.

There were many encouraging results in this analysis of charter captains' practices and attitudes, but we also found some cause for concern. Hammerhead sharks—which in Florida waters typically means great, smooth, and scalloped hammerheads, though there are other hammerhead species elsewhere—are highly valued by anglers for their "fight," and are accordingly targeted by charter captains. The problem is that these animals are extremely physiologically vulnerable to capture stress. This means that they often die even if you immediately release them—so no, angry anglers in my social media mentions, "he released it" does not mean "it survived." Florida's species of hammerheads are also, according to the IUCN Red List, Endangered or Critically Endangered, and are protected in Florida waters. Overall, though, charter captains in Florida are aware of and concerned by shark population declines and support efforts to protect shark populations.

I also studied land-based shark anglers, people who fish for sharks from beaches, piers, or bridges rather than from boats, during my doctoral research. This activity attracts a somewhat different socioeconomic group of anglers (fishing from a boat is expensive, and many of the self-described "average Joes" in the land-based fishing community can't afford to do this). This type of fishing is also rather poorly studied. It was relatively easy for me to track down everyone in Florida with a charter boat captain license, but land-based fishing is much more diffuse. People can do it anywhere, and until recently you didn't need a special license to fish, just the basic fishing license that millions of Floridians have. We knew that land-based fishing may cause extra physiological stress to sharks. Right before a shark is caught by a boat-based angler, it's in deep water, so it can breathe and its organs are supported. But on a shark's way to a land-based angler, it's dragged out of the water and over sand, concrete, or wood. It can't breathe during this time, and it may suffer abrasive injuries, including to the sensitive gills. Still, no one really knew how much of this riskier kind of fishing was happening, or which species were commonly targeted. To learn more, my colleagues and I used a technique called *digital ethnography*, analyzing thousands of postings on an online message board dedicated to land-based shark fishing. I'm pleased to report that our resulting paper has since been highlighted as an example of high standards of online research ethics for taking care to protect the privacy of individual anglers, even as we exposed problematic practices in the community as a whole.

As we analyzed their message board posts to get a sense of the land-based shark community's practices, we became quite alarmed. We found that lots of endangered, protected species are being illegally handled in ways that would likely lead to their deaths. We documented a minimum of 378 cases of unequivocally illegal fishing behavior involving protected species, along with thousands of borderline cases. Even more troublingly, we read statements from these anglers that clearly demonstrated that they knew they were breaking the law and didn't care. The boards even featured evasive scripts for club members to recite to law enforcement if they're caught breaking the law. Most of these anglers made it pretty clear that they don't have an especially high opinion of environmentalists or scientists. When my dad read my account of this experience, he thought someone calling me a "pompous pale-skinned bookworm" was the Funniest Thing Ever.

Looking at the big picture, recreational angling isn't the reason why shark species are threatened. Nonetheless, it's also true that many already threatened shark species are targeted by recreational angling. And vulnerable sharks aren't well-suited to withstanding additional threats.

Habitat Destruction

> The one process now going on that will take millions of years to correct is the loss of genetic and species diversity by the destruction of natural habitats. This is the folly our descendants are least likely to forgive us.
>
> ——E. O. Wilson

Typically, when people consider habitat destruction as a conservation threat, they think about cutting down the Amazon rain forest. How can a threat like this relate to animals who migrate as widely as sharks? Well, many species of sharks spend the first year or so of their lives in specific habitats called nursery areas, which are shallow and close to shore—and therefore close to humans. The big advantage of a nursery area for a small shark is that there's lots of food nearby. Being close to shore means lots of nutrients, in the form of small fish and invertebrates that prefer coastal waters. As a plus, there aren't a ton of larger (non-human) predators around these areas because the water is so shallow.

My master's research on sandbar sharks #BestShark focused on their role in the food web of a coastal nursery in South Carolina, an area that is thankfully protected from development. However, one of the most studied, most famous shark nursery areas in the world, that of Bimini in the Bahamas, is under threat due to mangrove removal.

Mangroves are a saltwater-tolerant plant whose complex root structure makes for lots of hiding places for small fishes. Newborn lemon sharks, which are common in the Bahamas, like to hang out in and around mangrove roots. But because mangroves live right on the border between land and sea, they're often removed to make room for construction projects that need seaside real estate. It's a major problem around the world, and it's threatening lemon sharks in Bimini.

Bimini is the site of the world-famous Shark Lab, as seen on more documentaries than even the Shark Lab staff can count. The doctoral research of Shark Lab scientist Dr. Kristine Stump took advantage of a Bimini construction project to provide some of the only evidence in the world of exactly what happens to a shark population and its associated prey community when a mangrove nursery area is destroyed. The results aren't pretty (as you can see in Plate 8 in the color insert): many lemon sharks died, it took the survivors longer to grow, and there were major disruptions to the whole food web. Even more troublingly, once a habitat was destroyed, the sharks stayed there rather than try to find a new, more suitable home elsewhere, essentially going down with the ship. (Habitat destruction like this also affects shark prey, of course, but the consequences of prey loss on shark conservation are not that well studied.)

Although a 2014 analysis of the top threats to sharks found just nine species which are in trouble specifically due to habitat destruction or degradation, several of these species are Critically Endangered. Eight of these nine species are threatened due to residential and commercial development. The remaining species, a freshwater shark, is threatened as a result of dam construction on a river. So, although overfishing is a greater threat to more species of sharks, habitat destruction is indeed a threat to some.

We can't fail to take into account the massive impacts of habitat destruction on ocean life as a whole, though, even if it is a bigger threat to many other marine organisms than it is to sharks. A common fishing method called *bottom trawling*, for instance, can be so destructive to the seafloor that it's been compared to burning down a whole forest to catch a few rabbits. While sustainable uses of trawl gear exist, there are lots of places with sensitive habitats where we simply should not trawl. Destroying mangroves to make shrimp aquaculture farms in Southeast Asia is an environmental catastrophe. Dynamite fishing (which is exactly what it sounds like) shocks my students when they hear about it. As with many of the threats to our oceans, how big a problem it is depends strongly on your focus: are you zeroing in on a particular group of species, or are you looking at marine biodiversity as a whole? We need to focus on conservation at all scales, narrow and broad, but we also need to be clear about what we are focusing on at a given moment.

Climate Change and Ocean Acidification

More CO_2 could lead to increased shark bites. Keep the sharks at bay. Take Metro.

—An ad for the Washington, DC, public transit network (which contains a "fact" about sharks that is notably incorrect and has been the bane of my existence for months)

Climate change is without a doubt one of the greatest environmental challenges of our time, a time filled with a great many environmental challenges. So you may be wondering why I have it categorized as a minor threat to sharks. Well, as surprising as this may seem, it is only a minor threat to sharks! Just one threatened shark species, the New Caledonia catshark (*Aulohalaelurus kanakorum*) is imperiled, at least in part, due to climate change. If we completely solve the problem of over-fishing but don't solve the problem of climate change, sharks will still be in good shape. However, if we completely solve climate change but don't solve overfishing, many species of sharks are still in big trouble. This doesn't mean that climate change isn't a worry or that we should ignore it, it just means that it isn't that much of a threat to sharks specifically. We absolutely should do everything possible to mitigate the worst effects of climate change, but we can't stop there, because climate change isn't the only threat to the environment.

Let's briefly discuss what I mean when I say that climate change just isn't really that big of a deal to sharks. In January 2020, I reviewed some of the ways that climate change and ocean acidification can influence shark biology and behavior for a blog post in which I ranted about the DC Metro ad mentioned above (for more detail on any of this, please read my *Southern Fried Science* post entitled "What the Hell Is DC Metro's 'Climate Change Will Increase Shark Bites' Ad Talking About? An Investigation." I hate that ad so much. It tasks me).

Anyway, the most common way that climate change affects marine life is simple: the water gets warmer, and animals that are adapted to

The author next to the offending Washington, DC, Metro ad. *Courtesy of Jacob Fenston, WAMU/NPR*

that particular temperature range are no longer well-suited for their changed environment. Animals then either struggle to survive (spending more energy on basic survival means having less energy to spend on growth, reproduction, and fulfilling their ecological roles), die, or move to an area that's now more suitable for their desired temperature range. Movement like this is called a *climate-induced range shift*, and it's happening to thousands of species all over the world on land and in the sea, which has the potential to be disastrous for not only wildlife but agriculture. Range shifts are even happening to sharks. I was a coauthor on the first paper to document this for a large, potentially dangerous species of shark along an inhabited coastline (Chuck Bangley, who you'll meet later in this book, was lead author on the paper).

We found that bull sharks, a species sometimes associated with human injury or death, once used nursery areas as far north as northern Florida. A coastal shark survey in the Outer Banks of North Carolina only very rarely saw any newborn bull sharks from the 1960s until 2011, although adults were spotted during this time period. Every year since then, though, the survey has caught multiple newborn bull sharks. This movement north correlates exactly with increasing water temperature in the Outer Banks. This habitat has become suitable for them when it obviously didn't used to be.

What does this mean for shark conservation and marine ecology? Is it bad for the marine ecosystem that a large predator has moved in? It sure could be, but we don't really know yet. Is this bad for humans? You may recall that there were a series of highly publicized shark bites in the Outer Banks in the mid 2010s where bull sharks were identified as likely perpetrators. This certainly doesn't imply that climate change makes shark bites more likely, since adult bull sharks were already in this area. Is climate change bad for bull sharks? I don't think so; if anything, their ability to expand their range shows they're pretty adaptable. A 2010 analysis found that most species of Great Barrier Reef sharks are not especially threatened by climate change because if the waters get too warm they can just move to a new habitat.

This idea of range shifts is one of the most likely sources of the DC Metro's very bad ad. I'm guessing that the ad's designer was inspired by an equally bad 2016 article in *TIME* magazine entitled "How Climate Change Is Fueling a Rise in Shark Attacks" which focused on climate-induced range shifts. It noted a (slight) increase in shark bites from 2015 to 2016 and argued that climate change may be prompting sharks to move into new habitats. However, it failed to point out that the shark species responsible for biting humans in 2016 were not those identified as having experienced climate-induced range shifts.

Climate change also causes ocean acidification, which is a complex concept to explain. Essentially, when there's more CO_2 in the atmosphere it leads to more carbonic acid in the ocean, which lowers the pH enough to disrupt all kinds of biochemical processes in marine life. Sometimes it can even dissolve the shells of marine animals. It's nasty stuff. In sharks, ocean acidification may cause birth defects, including dermal denticle formation or stunted growth. These defects make it harder for them to smell prey or disrupt their normal behavior in other ways. Sharks in acidified waters, for example, may exhibit increased boldness or activity levels, perhaps because acidification disrupts their sense of smell. It may also affect sharks indirectly; the destruction of coral reefs and their associated prey is bad news for sharks that feed on reefs (though many of them are not *only* found on coral reefs). These problems are certainly not great for sharks, but as threats they pale in comparison to the overfishing that has killed thousands of sharks just during the time you've been reading this chapter.

Plastic Pollution

I want to say one word to you. Just one word. Are you listening? Plastics.

—Mr. McGuire (Walter Brooke), *The Graduate* (currently ranked as one of the American Film Institute's top 100 movie quotes of all time)

In the last few years, plastic pollution has become one of the best-known ocean conservation problems. Public awareness of the issue has been driven by heartbreaking imagery of dead whales and sea turtles. Although marine plastic pollution is a huge ocean conservation problem facing many species and ecosystems, it is not a population-level threat to sharks. In other words, as we found earlier when examining the threat posed by climate change, if we totally solve overfishing and don't solve the plastic pollution problem at all, sharks will be in great shape. However, if we completely eliminate plastic pollution and don't put a stop to overfishing, sharks are still in pretty big trouble. We should work on solving the problem of marine plastic pollution because it's a serious threat to many animals and ecosystems, but it's not really a shark conservation concern. Ultimately, it harms sharks even less than climate change.

Whenever I mention that plastic doesn't threaten sharks in my "Ask Me Anything" sessions on social media, it results in lots of follow-up questions, so I'd like to briefly explain how we know that plastic pollution just isn't that big of a deal for sharks. Plastic is a problem for marine life for two main reasons: entanglement (they get physically stuck in it, which immobilizes them or causes some kind of sublethal injury) or ingestion (they eat it and it poisons them or blocks their airway or digestive tract).

Do sharks get entangled in marine debris? Absolutely yes, they do. A 2019 paper led by UK marine scientist Kristian Parton identified about

600 documented cases of 34 shark and ray species getting entangled in abandoned fishing gear. In about 74% of cases of entanglement in the scientific literature, sharks and rays were harmed by fishing gear; the remaining 26% of the time, they became entangled in things like plastic straps, bags, packaging, and even tires. Three hundred thirty-four of these animals were totally trapped, while others were partially entangled or had a piece of plastic debris stuck to them. If we assume that all of these animals died (and not all of them did), that means that more sharks were killed by overfishing while you were reading this page than have ever been confirmed killed by plastic debris entanglement.*

Do sharks ingest plastic debris? They definitely do. In another study led by Kristian Parton, the authors found that more than two-thirds of all the shark stomachs and intestines they analyzed contained at least two pieces of microplastic. One of these sharks had nearly 150 pieces of plastic in its stomach. So there's no doubt that when microplastics are present in the environment, predators end up with microplastics in their stomachs. Is that toxic to them? It probably isn't fantastic, and such suffering is certainly an animal welfare issue, but this study points out that the long-term health effects are currently unknown. Do microplastics end up blocking sharks' digestive tracts? They don't seem to. For what it's worth, the lead author of both these papers, Kristian Parton, wrote in his dissertation that "these two threats are unlikely to have significant population impacts on sharks and rays globally," so I'm not editorializing here. The same papers that show that sharks do get entangled and do ingest plastic also say that, nonetheless, plastic doesn't represent a population-level threat to them.

What about filter feeding sharks, which swim around with their mouths open? A 2018 paper led by Elitza Germanov noted that whale sharks and manta rays filter feed in areas with lots of microplastic pollution in the water, and are very likely ingesting microplastics. However, this same paper pointed out that there is no evidence that plastic pollution is a population-level threat to these animals.

The best available evidence is pretty clear here: plastic pollution, which

*I'm assuming here it took you around four minutes to read this page, but maybe some of you are speed readers! At any rate, you get the point.

is a big problem for lots of reasons that we should absolutely work to solve, is just not a major shark conservation concern. (That doesn't mean that there's no chance that plastic pollution can be a population-level threat in the future. If our goal is evidence-based policy and triaging limited resources, though, we have to consider that we have lots of evidence that some threats are serious to sharks, but currently no evidence that this one is especially serious to sharks.) I should also mention that *other types* of ocean pollution can be harmful to sharks, disrupting reproductive systems or even killing individuals. These, however, don't typically lead to population-level threats. While I have your attention, there are some other important public misunderstandings about plastic pollution, why it's a problem, and what the solutions to it are. These don't relate to sharks so I'll be brief, but they are important.

First of all, the Great Pacific Garbage Patch, which is often invoked in discussions about the need for plastic pollution controls, is not what you (probably) think it is. It's often visualized as a huge island of plastic trash dense enough to walk on. Many people also assume that marine life gets stuck in it. In reality, the Patch is more like a soup of tiny plastic particles; check out the video of it on this book's website if you're interested in seeing how it really looks.

Related to this misconception, you may have heard of the Ocean Cleanup, a nonprofit environmental engineering organization that is trying to eliminate marine plastic pollution. It has built several systems to deal with plastic marine debris. Often promoted as brilliant solutions to the marine plastic pollution crisis, each of these systems is essentially a big net floating in the ocean that aims to scoop all the plastic out. My longtime social media followers know that there's enough wrong with this idea to write a whole other book, but in short, the Ocean Cleanup systems demonstrate major misunderstandings of the problem with ocean plastic pollution. Net-based solutions are very likely to do more harm than good to marine life; we've already seen that huge nets result in huge bycatch. Remarkably, the Ocean Cleanup's own promotional materials featuring the first test run of their system show hundreds of dead animals caught in their net, yet this is advertised as a success story somehow. I haven't found a single expert in marine plastic pollution who thinks this is a good idea, and many I've spoken to correctly predicted exactly how it would eventually fail. This is one of my go-to case studies for why it's important to listen to experts.

Wildlife Tourism

Some readers may find it odd to see wildlife tourism listed under threats to sharks, since it's often (wrongly) presented as the best or only solution to protecting endangered species. Sure, wildlife tourism can be a part of the solution in some cases, as we'll discuss later in this book. But keep in mind that there are many types of wildlife tourism operators. Some run their businesses in ways that are safe for humans and sharks while promoting responsible stewardship of marine resources. Unfortunately, other operators are a type that I classify using the technical term "macho cowboy idiot." Please note that "macho cowboy idiot" is a gender-neutral term; some of the worst offenders are women. Some wildlife tourism operations absolutely cause harm to wild animals. The really irresponsible ones tend to artificially crowd many sharks together during mealtimes, resulting in increased animal stress and disease.*

Defenders of the worst practices of macho cowboy idiot dive operators say that causing sharks to become stressed, diseased, and/or injured is better than allowing them to be killed to make shark fin soup. This might be a reasonable argument if these were the only two possible options.

Are Aquariums a Threat?

"I'd do it again," says B.C. man who jumped naked into Toronto shark tank.

—Liam Casey, article headline for *The Canadian Press*

Since the release of the anti-captivity documentary *Blackfish*, I've been asked a lot about whether it's bad to keep sharks in captivity. World-

---

*My younger brother went diving with whale sharks at Oslob in the Philippines on his honeymoon. Afterward, he called me and said it was the coolest thing he'd ever done, but that he felt horrible the whole time because there were just too many people there.

wide, the number of sharks in captivity pales beside the number of sharks killed by commercial fisheries in a day, which indicates that this just isn't a population level problem. In a lot of cases, exhibiting sharks in zoos and aquariums isn't even really an animal welfare issue. There are some concerns about keeping certain marine mammals in captivity, but for the most part they don't really apply to fish, which have notably less well-developed brains and social structures. Most sharks can be totally fine in captivity. As long as they're kept in safe conditions, this doesn't bother me at all, even if you don't factor in the enormous good that aquariums do for research, conservation, and public understanding of ocean issues.

As a diverse and wide-ranging group of animals found in many habitats around the world, sharks face many kinds of threats. (As you now know, this issue goes far beyond shark finning.) A detailed understanding of the scope of the conservation crisis facing sharks, along with the multiple reasons they are threatened, is essential for anyone who hopes to help these animals recover.

In this chapter, you've seen that many misconceptions persist about which specific threats to sharks are most harmful, as well as what the possible effective or actionable solutions to those threats might be. As you continue to learn more about sharks, I encourage you to seek out reliable expert sources rather than listening to the loudest fringe voices you can find on social media. A good general rule is that if someone is presenting a complex worldwide problem as extremely easy to solve, they're not telling you the truth, or at least not the whole truth.

5 » How Can We Protect Sharks?

Which Species Should We Conserve and Protect?

Don't boo, vote!

—President Barack Obama

By this point in the book, you know a lot about sharks. You know that sharks are amazing animals because of their unique biology and complex, fascinating behaviors. You know that there are hundreds of shark species, including some of the weirdest and coolest fish in the ocean. You know that despite "If it bleeds, it leads" media hysterics, sharks aren't a significant threat to people. You know that sharks play a critical role in their ecosystems, which means that humans are much better off with healthy shark populations than we are without them. And in the last chapter, you heard the bad news that many shark species are in big trouble as a result of our own activities.

Now I have some positive news: the problem of declining shark populations, while daunting in scale and potentially devastating in impact, is solvable. In many cases, we already know what needs to be done and, even better, we know how to do it.

There is some important background information that you need to understand in order to make sense of our available options to help sharks recover, including a variety of different conservation biology theories that can be used to assist us in prioritizing which of the many shark species to protect. Numerous types of conservation and management policies can be used to protect sharks; this chapter summarizes the scientific evidence showing how they work, as well as when (or if) they work. I'll next attempt to explain the various facets of these policy options fairly, as well as the arguments for or against various policies, while noting that I often personally feel pretty strongly about which ones are more effective.

I'll do my best to summarize these philosophies here, and connect them to our favorite toothy subject. You'll notice more appearances of our tour's recurring theme: a lot of the information about sharks that you've seen before is . . . not entirely correct. First of all, no, we cannot just conserve every single species. Not only do we not have the resources to do that, but from a conservation science perspective it isn't actually necessary. Not every species of shark needs our attention; many are doing just fine. With limited personnel, funding, and political capital, we have to pick and choose what to protect. There are a few different ways that conservation scientists and environmental activists select which species to focus their efforts on.

#1. Protect the most charismatic species because people love charismatic species, so it's relatively easy to generate public support and raise funds

> For too long, the cute and cuddly creatures have ruled the roost!
>
> —Simon Watt, founder, Ugly Animal Preservation Society,
> author of *The Ugly Animals: We Can't All Be Pandas*

Everyone loves adorable animals with fuzzy paws and soulful eyes, so it's no coincidence that one of the largest and most successful environmental nonprofit groups in the world, the World Wildlife Fund, uses a panda as its logo and mascot. It's easier to get people emotionally engaged with something charming, but it's harder to get them to care about snakes, spiders, or sharks. A particularly powerful World Wildlife Fund ad shown below drew attention to this, asking if people would care more about, for example, endangered bluefin tuna if they sported furry, endearing mammal faces. (Incidentally, with standard "Don't try this at home" disclaimers in mind, Steve "The Crocodile Hunter" Irwin was excellent at getting people to care about species that aren't cute and cuddly. I think we have a lot to learn from his approach.)

Sharks are hardly adorable. This was explicitly mentioned in a paper that identified charismatic animals by surveying members of the public, looking at featured species on zoo websites and signage, and examining which animal species are featured in Disney films. Of the 20 species judged to be "charismatic," 18 were mammals, and 16 of those primarily live on land. Surprisingly, the great white shark was listed among the usual suspects, including the perennially popular lion, elephant, and yes, panda. However, the research team also had their survey subjects decide which adjectives best fit each of the top 20 species, with choices including beautiful, endangered, impressive, dangerous, rare, and cute. Nineteen different species were described by all six of these words at least once, but the great white shark was never even once described as cute by anyone. So we have to use a modified version of the charismatic species concept for sharks, one focused on them as powerful, awe-inspiring, cool, or even scary animals. This can work, but it doesn't convince everyone.

Since the 1980s, popular, charismatic species have also been known

This ocean conservation campaign poster demonstrates that fish (here, tuna) don't get the same amount of love from the public as cute, furry mammals. *Courtesy of the World Wildlife Fund*

to conservationists as *flagship species*. The idea is that well-liked species can serve as symbols for conservation in ways that benefit less famous or less beloved species, whether because they belong to the same ecosystem or geographic territory as a more popular animal or because they, too, could benefit from an increase in conservation donations. Utilizing a flagship species in a marketing campaign or fundraising effort can absolutely work for sharks sometimes, though the threats facing the most charismatic species aren't always the same as the threats facing the species in the most dire straits.

#2. Protect the species most threatened with extinction

> It's easy to think that as a result of the extinction of the dodo, we are sadder and wiser, but there's a lot of evidence to suggest that we are merely sadder and better informed.
>
> —Douglas Adams and Mark Carwardine, *Last Chance to See*

A particularly straightforward way of prioritizing conservation efforts and effectively directing limited resources is to focus on the species in the greatest need of conservation assistance—those that are the most threatened, meaning those that might go extinct if we don't act now. This is often referred to as *conservation triage.*

The IUCN Red List is run by the IUCN Species Survival Commission, which is an international, apolitical group of about 10,000 volunteer scientific experts divided into approximately 160 specialist groups and task forces. They systematically evaluate how much conservation trouble the world's fungi, plants, and animals are in. They have created different categories, each of which has strict, technical, consistent definitions across species and regions, so we can easily, replicably, and authoritatively identify which species are in the worst shape.

The criteria guide fits on one page (I have it hanging on my office wall). But the guidelines on how to apply these criteria span more than 160 pages, not counting large technical documents of mapping guidelines and regional application guides. A common misconception among experts who haven't completed IUCN Red List training is that there's some degree of opinion involved in determining which species get assessed as falling into which category. In reality, not only are the category definitions quite rigorous but the assessments that the group relies on are evidence based and data driven. The IUCN Species Survival Commission has a Shark Specialist Group (SSG) made up of about 200 experts from all over the world. As of this writing, I work with the IUCN's Tuna and Billfish Specialist Group, and you'll see my name attached to the conservation status of dozens of species of tuna and their relatives.

The Red List categories include one called Not Evaluated, which lists species that IUCN experts simply haven't looked at yet. Among those species that *are* evaluated, there is a categorization system of increasing conservation risk. In Plate 9 in the color insert, the closer to the top of the diagram the species assessment is, the worse shape they're in.

Even if a species is evaluated on the Red List, it can be considered Data Deficient. Species that fall into this category have been assessed by experts, but there simply isn't enough scientific data to make a definitive determination about how threatened they are. Some people seem to wrongly believe that Data Deficient means "We don't know anything at all about this species, so anything we can learn is useful for their conservation." That is simply not what the term implies. The data we're looking for when we evaluate species' conservation risks must relate to population trends over time; random discoveries about a species aren't useful for making an assessment. The Red List's technical guide to categories and criteria explains that the Data Deficient designation means that "a taxon in this category may be well studied, and its biology well known, but appropriate data on abundance and/or distribution are lacking."

The lack of scientific data about the population of a species is often, but not always, a bad sign in terms of how healthy a population is. If there are a lot of individuals left and the species is healthy, it's easier to get data about the status of their population than if the species is rarely seen because their population is low. At the end of the last comprehensive assessment in 2014, there were 487 Data Deficient species of sharks and their relatives, including extremely well-studied species like the horn shark, *H. francisci*. We study these animals all the time (in fact, a Scopus library database search for horn sharks shows that there have been 82 papers published that mention this species), but we don't have data on their population trends.

A 2014 paper featuring dozens of Shark Specialist Group experts from around the world attempted to estimate the status of Data Deficient shark species. It's hard to make science-informed conservation policy for a species that is assessed as Data Deficient, so they tried to assess whether it is possible, using scientific models, to make informed estimates of what each species is likely to be, based on what we know about them. From the species that could be assessed, the authors noted that bigger species that live in shallower water and can only tolerate a narrow range of depths tend to be more threatened. They argue that, because we have some of that data for many of the Data Deficient species (as well as expert knowledge that can be used for inference), we can estimate how threatened they're likely to be.

When there is enough data to evaluate a species, the best-case scenario is an evaluation of Least Concern. This means that the species is doing fine. It also implies that it probably shouldn't be a major focus of conservation because there are many species doing a lot worse that could really use the help. The Red List criteria guide refers to these species as Widespread and Abundant. Sharks listed as Least Concern include the coldwater salmon shark, whose pups are sometimes confused with great whites along Pacific Northwest beaches. Notably, some species have specific populations that are doing well, even if the species is doing poorly overall. One example is the bonnethead shark, a small member of the hammerhead family commonly caught by anglers in Florida. Bonnetheads are assessed as Least Concern in the United States, but are Endangered elsewhere.

If a species doesn't meet any of the technical criteria to be assessed under a threatened category, but is clearly not Least Concern, it will be assessed as Near Threatened. These species are not doing particularly well, but are not in nearly as much trouble as those in a threatened category. A Red List colleague once told me that he wished that people would keep in mind that Near Threatened means *not* threatened. However, the Red List guide points out that such species are "close to qualifying for, or are likely to qualify for, a threatened category in the near future."

These are species to keep an eye on. For example, the well-known tiger shark was reassessed as Near Threatened in 2019. The assessment states that, while its population numbers haven't quite declined enough for tiger sharks to officially rate as Vulnerable, that threshold is almost met and the threats are likely to continue.

In increasing order of extinction risk, the remaining Red List categories—Vulnerable, Endangered, and Critically Endangered—are collectively referred to as the *threatened* categories. Observe that the actual categories are capitalized, while the group of threatened categories is not. (Missing this difference in capitalization is probably the most common thing I've ever dinged someone for during peer review of their scientific manuscript.) To determine if a species belongs in one of the threatened categories, there are five technical criteria that each species is assessed against:

A. population size reduction, which is measured over 10 years or 3 generations, whichever is longer,

B. geographic range,

C. small population size *and* declining population,

D. very small or restricted population, and

E. quantitative analysis of extinction probability.

Each has its own subcriteria. Many of these were designed for terrestrial conservation and don't make a lot of sense in the ocean, so most shark assessments only deal with criterion A.

For example, a recent high-profile addition to the list of sharks classified as Endangered is the shortfin mako, which now qualifies for the category under Criteria A2bd. Most people only talk about the overarching category of Endangered, but if you're familiar with Red List terminology there's a lot of interesting information in the rest of the assessment. For example, the mako's A2 criterion here refers to a situation where "population reduction [has been] observed, inferred, or suspected in the past, where the causes of reduction may not have ceased, or may not be understood, or may not be reversible." This means that mako shark numbers are declining at an alarming rate, and the cause of their decline is still out there—so they need help fast. Subcriteria A2 contains 5 different possible cases, of which mako sharks meet 2 (b and d). A2b means that 2 is the case based on "an index of abundance appropriate to the taxon," and A2d means 2 is the case based on "actual or potential levels of exploitation." These refer to what type of data is available about the existing threats and how severe they are, as well as how confident we are about the problems these threats pose.

Under Criterion A2, Endangered means that there's been a greater than 50% but less than 80% population reduction over the last 10 years or 3 generations. This all gets technical very quickly, but the important point here is that Red List assessments are based on science and data and are not based on someone's opinion or guesswork. They are also independently replicable. In other words, anyone applying the same criteria to the same data should come to the same conclusions. They don't have the force of law—an IUCN Red List assessment of Endangered is not the same thing as an Endangered determination under the United States Endangered Species Act, for example, but they mean something real and important.

So how many species of sharks are in trouble? According to that 2014 paper led by Nick Dulvy at Simon Fraser University, 24%—nearly a full quarter—of all known species of sharks and their relatives are assessed (or estimated, if they're Data Deficient) as belonging to one of the threatened categories. Currently, amphibians are in worse shape because of a global fungal disease outbreak that's killing frogs, but few other groups of animals are as threatened as sharks. Even more concerningly, sharks and their relatives have fewer species assessed as Least Concern than any other group of animals in the world.

The species of sharks listed as Endangered and Critically Endangered can be found in the table on page 106, and is correct as of summer 2019. This list includes a lot of species that you've probably never heard of, but does *not* include the most famous shark species, the great white, which is assessed as Vulnerable. Although great white populations are struggling, they're absolutely not one of the species in the most dire need of attention to avoid extinction. In 2020, I conducted a survey of North American shark conservation experts in which I asked people to name species that need more attention. One respondent replied, "Dear God, not great whites!"

Don't blame yourself, by the way, if you haven't heard of some of these species; it's a problematic fact that Critically Endangered shark species are rarely mentioned in popular press coverage of shark conservation. In my own research, an analysis of over 1,800 mainstream media articles about shark conservation found only 20 mentions of any Critically Endangered species. Part of the problem is the enormous amount of attention taken up by non-endangered species. For example, there is, as of this writing, a Facebook group for non-expert shark enthusiasts called White Shark Advocacy that has over 7,000 members. It explicitly states that its only purpose is to talk about protecting great white sharks; the group's moderators mention that they will delete posts about other topics and other species. A similarly narrow group, White Shark Interest Group, boasts over 34,000 members. There is not a single Facebook group dedicated to protecting Critically Endangered species like the striped smooth-hound or the daggernose shark.

This is a shame because some of these lesser-known species are awesome. The aptly named daggernose's snout is even pointier than a swordfish's, a stark visual contrast with its big, flat, paddle-shaped fins. The sawback angelshark sports punk rock spikes protruding from its spine. The Natal shyshark is called "shy" because it covers its face with its tail to hide. Contrary to what most people believe (and what Shark Week programming has alleged) bull sharks are absolutely *not* the only shark species that can survive in fresh water: sharks of the genus *Glyphis* also live most of their lives in it. And the winghead shark, a hammerhead relative, has one of the craziest heads I've ever seen on a fish.

A winghead shark, one of nature's delightful weirdos. Seriously, look at this thing. *Courtesy of Wildscreen Arkive/Steve Kajiura*

Table 5.1 Shark species assessed as Endangered or Critically Endangered by the IUCN Red List as of 2021

Species	Scientific Name	Red List Assessment	Criteria
Pondicherry shark	*Carcharhinus hemiodon*	Critically Endangered	A2acd; C2a(i)
Ganges shark	*Glyphis gangeticus*	Critically Endangered	A2cde; C2b
Northern river shark	*Glyphis garricki*	Critically Endangered	C2a(i)
African spotted catshark	*Holohalaelurus punctatus*	Critically Endangered	A2abcd+3bcd+4abcd
Striped smooth-hound	*Mustelus fasciatus*	Critically Endangered	A2abd+3bd+4abd
Smoothback angelshark	*Squatina oculata*	Critically Endangered	A2bcd+3cd
Angel shark	*Squatina squatina*	Critically Endangered	A2bcd
Irrawaddy river shark	*Glyphis siamensis*	Critically Endangered	B1ab(iii,v)
Natal shyshark	*Haploblepharus kistnasamyi*	Critically Endangered	B1ab(iii)
Daggernose shark	*Isogomphodon oxyrhynchus*	Critically Endangered	A2ad+3d+4ad
Sawback angelshark	*Squatina aculeate*	Critically Endangered	A2bcd+3cd
Argentine angelshark	*Squatina argentina*	Critically Endangered	A2bd
Hidden angelshark	*Squatina occulta*	Critically Endangered	A2bd
Borneo shark	*Carcharhinus borneensis*	Endangered	C2a(ii)
Whitecheek shark	*Carcharhinus dussumieri*	Endangered	A2d+3d
Speartooth shark	*Glyphis glyphis*	Endangered	C2a(i)
Whale shark	*Rhincodon typus*	Endangered	A2bd+4bd

Smoothtooth blacktip shark	*Carcharhinus leiodon*	Endangered	A2cd+3cd
Harrisson's dogfish	*Centrophorus harrissoni*	Endangered	A2bd
Winghead	*Eusphyra blochii*	Endangered	A2d+3d
Whitefin topeshark	*Hemitriakis leucoperiptera*	Endangered	B1ab(iii,v); C2a(ii)
Honeycomb izak	*Holohalaelurus favus*	Endangered	A2abcd+3bcd+4abcd
Shortfin mako	*Isurus oxyrinchus*	Endangered	A2bd
Longfin mako	*Isurus paucus*	Endangered	A2d
Broadfin shark	*Lamiopsis temminckii*	Endangered	A2d+3d
Narrownose smoothhound	*Mustelus schmitti*	Endangered	A2bd+3bd+4bd
Scalloped hammerhead shark	*Sphyrna lewini*	Endangered	A2bd+4bd
Great hammerhead shark	*Sphyrna mokarran*	Endangered	A2bd+4bd
Greeneye spurdog	*Squalus chloroculus*	Endangered	A2bd
Taiwan angelshark	*Squatina Formosa*	Endangered	A2d+4d
Angular angelshark	*Squatina guggenheim*	Endangered	A2BD
Zebra shark	*Stegostoma fasciatum*	Endangered	A2bd+3bd
Sharpfin houndshark	*Triakis acutipinna*	Endangered	B1ab(v); C2a(ii)

Although this book focuses only on sharks, many closely related rays are in pretty sorry conservation shape, too. The sawfish, also known as the carpenter shark, is actually a large ray. It was considered the most threatened marine fish in the world until recently, when that dubious honor was given to the guitarfish and wedgefish, which are also rays.

(Incidentally, if you are interested in learning to assess the conservation status of different species, the IUCN Red List training course is free online at www.conservationtraining.org. It took me about a week to complete the course and pass the online exam.)

#3. Protect the most evolutionarily unique species

> Don't worry about not fitting in. The things that make people think you're weird are what makes you you, and therefore your greatest strength.

—Birgitte Hjort Sørensen

If we have to pick and choose which species we're going to protect—and I cannot stress enough that we do—another way to accomplish this is by highlighting the species that don't have any close relatives. The theory here is basically that it's worse for biodiversity if a unique species goes extinct than if a species with many close relatives goes extinct.

To this end, the Zoological Society of London runs a program called EDGE of Existence, which identifies evolutionarily distinct and globally endangered species. As the EDGE program's website puts it, "Some species are more distinct than others because they represent a larger amount of unique evolution. Species like the platypus have few close relatives and have been evolving independently for many millions of years. Others, like the brown rat, originated relatively recently and have many close relatives." This doesn't mean that it wouldn't be problematic for a less distinctly evolved species to go extinct; extinction is always bad. It instead highlights that something extra is lost when a species with no close relatives goes extinct, and there's value in trying to avoid that.

In 2018, a research team that included several of my former labmates applied the EDGE criteria to sharks and their relatives. Their resulting paper identified several sawfish and guitarfish, as well as other rays and skates, but in keeping with this book's theme, I'm going to focus only on the sharks here. The EDGE shark species can be found in the table on page 109. It includes some of the true weirdos of the shark world, like the aforementioned winghead, the basking shark (which, like

Table 5.2 The sharks on the EDGE top 50 list of sharks and their relatives, including a Zoological Society of London (ZSL) assessment of how much conservation is currently taking place around this species

EDGE rank (1 highest, 50 lowest)	Species	Scientific Name	Conservation Attention
5	Angelshark	*Squatina squatina*	Good
6	Sawback angelshark	*Squatina aculeata*	Good
13	Smoothback angelshark	*Squatina oculata*	Low
15	Natal shyshark	*Haploblepharus kistnasamyi*	Very low
16	Winghead	*Eusphyra blochii*	Medium
17	Whale shark	*Rhincodon typus*	Good
18	Zebra shark	*Stegostoma fasciatum*	Low
19	Sand tiger shark	*Carcharias taurus*	Good
20	Bigeye thresher shark	*Alopias superciliosus*	Good
22	Great hammerhead	*Sphyrna mokarran*	Low
23	Basking shark	*Cetorhinus maxium*	Good
24	Bluegray carpetshark	*Brachaelurus colcoughi*	Very low
26	Snaggletooth shark	*Hemipristis elongata*	Low
29	Scalloped hammerhead	*Sphyrna lewini*	Low
32	Harrisson's dogfish	*Centrophorus harrissoni*	Medium
32*	Sharpfin houndshark	*Triakis acutipinna*	Very low
35	Whitespotted izak	*Holohalaelurus punctatus*	Very low
36	Honeycomb izak	*Holohalaelurus favus*	Very low
39	Great white shark	*Carcharodon carcharias*	Good
44	Kitefin shark	*Dalatias licha*	Very low
45	Tope shark	*Galeorhinus galeus*	Low
46	Porbeagle shark	*Lamna nasus*	Medium
47	Shortfin mako	*Isurus oxyrinchus*	Good
47*	Longfin mako	*Isurus paucus*	Low
50	Pelagic thresher	*Alopias pelagicus*	Low

Note: * = this is not a typo; some species have the same rank.

whale sharks, is a filter feeder; despite its massive size, it only eats tiny plankton), and the deep-sea kitefin shark (which, as was discovered while I was writing this chapter, is the largest bioluminescent fish in the world).

#4. Protect the most ecologically important species: the keystone species concept

All animals are equal, but some are more equal than others.

—George Orwell, *Animal Farm*

This is where the keystone species concept comes into play. The idea here is that there are species so ecologically important that losing them would be much worse for the whole ecosystem than you'd suspect based just on the size of their population. This conservation paradigm holds that protecting these species should be a prime concern, even if they aren't the most endangered, or the most evolutionarily unique, or the most charismatic. Many conservation advocacy efforts highlight the ecological importance of sharks as a reason to conserve them. My own research found that this is by far the most common reason why environmental advocates from nongovernmental organizations (NGOs) say that we should protect sharks—because sharks are important to a healthy functioning ecosystem, and the decline or loss of sharks can cause all kinds of nasty ecological disruptions.

There is indeed some evidence (depending on what ecosystem is being examined) from the world of ecological modeling that sharks can sometimes be keystone species, and conservation advocates regularly imply that they are keystone predators. Researchers have noted that in many models of marine ecosystems, sharks and rays have high levels of keystoneness, a quantifiable metric based on their relative *biomass* (a measure of the total mass of organisms in an ecosystem) and the strength of their ecological interactions with other species. For example, a 2004 ECOSIM/ECOPATH model of the wildlife surrounding Floreana reef in the Galapagos found that sharks had the third-highest keystone index

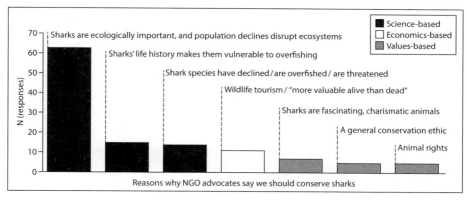

Reasons why NGO advocates say we should conserve sharks. *Courtesy of the author*

score of 43 tested species or species groups in that ecosystem. Only sea-birds and marine mammals had more ecological impact relative to their biomass. However, a 2002 study in the central Pacific Ocean found that sharks were not keystone predators of that particular ecosystem, primarily due to the presence of so many other ecologically similar predators in the area, like big tunas and billfishes. The high concentration of these predators constrained sharks from playing a disproportionate ecological role in that case.

#5. Protect species whose protection would also benefit nearby species: the umbrella species concept

The *umbrella species* concept in conservation biology suggests that we should focus on protecting species that would de facto necessitate protecting lots of other nearby species: those that fall under their "umbrella" of protection. A classic example of an umbrella species is the northern spotted owl, which needs an old growth forest habitat in order to survive. Protecting these owls, a charismatic species that people love, therefore requires protecting old growth forest habitat. This in turn also protects many other species that live in these forests but aren't necessarily as beloved (or easy to get public support for) as owls. Similarly, protecting the high profile (if not necessarily adorable) sage grouse also protects other animals that live in sage grouse habitat.

The umbrella species concept is more commonly applied to terrestrial conservation, at least explicitly, but it's often implicitly present in conversations about the ocean. Protecting the habitat of a highly migratory ocean animal like a large shark can benefit everything else that lives in that habitat, something that is often discussed as a side benefit of shark conservation without necessarily using the term "umbrella species." The umbrella species concept has been explicitly applied in marine ecosystems where the endangered humphead wrasse dwells. Providing protections for these fish helped to protect a whole coral reef. Researchers have suggested that scalloped hammerhead sharks could be a useful umbrella species for conserving the biodiversity surrounding *seamounts*, or underwater mountains that serve as oases in the desert of the open ocean. Whale sharks have been referred to as an umbrella species, and indeed Mexico's Whale Shark Biosphere Reserve (Reserva de la Biosfera Tiburón Ballena), created in 2009, was designed to save whale sharks by protecting their habitat. In the process, it also shielded other species that live in that habitat, including sea turtles and marine mammals.

So Which Conservation Philosophy Should We Use?

Which conservation philosophy is best? In reality, it depends on the situation. They can all be useful for some threatened species in some places in some contexts. Generally speaking, we should use the argument that has the highest chance of a successful conservation outcome for that specific situation, and in practice that's not always going to be the same one. In other words, conservation is complicated, details matter, and there's no one-size-fits-all or silver bullet approach that's always going to work.

Who Is Empowered to Protect Which Species?
A Note on Jurisdictions

Jurisdictional issues may not be the flashiest part of conservation, but you can't help threatened species if you don't know who has the power to help them. Details, knowledge, and expertise matter here; thinking outside the box can be great, but it helps if you know where the borders of the box are and why they are there in the first place.

I once saw an amateur online petition asking the United Nations to change Florida's shark fishing regulations. The United Nations has absolutely no authority to do that, which means that the petition in question cannot possibly have accomplished its goals. It got tens of thousands of signatures despite being totally useless in terms of policy change.

Some of these regulatory powers vary pretty significantly among nations. For example, in Canada, the federal Fisheries and Oceans Canada (DFO) agency sets all fisheries quotas without input from the country's provinces and territories, even in coastal waters. The domestic examples in this chapter come from the United States, where I live and work. I'll start at the smallest scale by discussing rules at the local/municipal level and work outward until we get to international and even worldwide rules.

Local/Municipal Rules

Very little fisheries management happens at the local or municipal level. Some recreational fishing restrictions are imposed at the county, pier, or marina level, however.

Right when I first started in the world of online public science engagement, there was a much-hyped (but ultimately short-lived) program called the Shark Free Marinas Initiative (SFMI). This initiative aimed to ban the *landing* (bringing sharks from a boat onto shore) of sharks at specific individual marinas. As a 2011 press release announcing the program noted,

The SFMI is a project of the Humane Society of the United States that enlists the support of fishermen, marinas and businesses in prohibiting and discouraging the recreational killing of sharks . . . hundreds of thousands of sharks [are] killed for sport in the U.S. annually. This recreational killing of sharks adds not only to the alarming death toll but also undermines efforts to gain better protections for sharks and sends an unspoken message that the only good shark is a dead one . . . SFMI marinas participate in a voluntary program where they agree to prohibit sharks from being landed at their facility.

The goal was to stop individual marinas (not whole cities, not whole states, not whole countries) from allowing boats that dock there to kill sharks. At one point, the Humane Society was aiming to get 1,500 marinas to enact this policy, but as of this writing, the www.sharkfree-marinas.com website is dead, and I haven't heard anything about this program in at least five years.

Some individual beaches or fishing piers also ban shark fishing. For example, following a series of shark bites in the North Carolina Outer Banks, Bogue Inlet, a fishing pier in Emerald Isle, announced a ban on shark fishing. The concern here was that putting shark bait in the water at the pier attracted sharks to the nearby beach, resulting in tourists being bitten by sharks which would not otherwise have been in the immediate area (at least not in feeding mode). Indeed, at least one shark bite occurred when a shark actively hooked by an angler tried to escape and encountered a swimmer while "fighting" the angler. For their part, anglers say they go to piers or sometimes beaches because sharks are *already* there, not in order to attract them from far away. Even so, we probably shouldn't be dumping large amounts of chum in the water right next to an area where a bunch of people are swimming.

State-Level Rules

This law says it's ok for the rich who can afford a big boat and gas bill to compete in kill tournaments. The poor ol' normal man who can't get past state waters is not allowed to play without harassment. As usual the rich get special [privileges] that the normal person does not.

—An anonymous recreational angler who fishes from a beach, as reported in a 2017 paper I coauthored, "Fishing Practices and Representations of Shark Conservation Issues among Users of a Land-Based Shark Angling Online Forum"

In the United States, individual states have jurisdiction in state waters. The federal government has jurisdiction throughout the rest of the country's *exclusive economic zone*, or EEZ, which extends out to 200 nautical miles from shore. State waters, for their part, extend from the shoreline to three miles out, except in the Gulf Coast of Florida, where they extend nine miles out. (Why? Because Florida.)

For commercial fisheries that exist entirely within the waters of one state, fishers are governed by rules and regulations set by that state's natural resource management agency, with names that vary between states. Fishers need to obtain a permit from that agency to fish. In the United States, most shark fisheries are not managed exclusively at the state level because the majority of shark stocks span state boundaries or are found mostly in federal waters. But a few are, or at least partially are. For example, in Louisiana state waters, commercial fisheries are allowed to catch 45 large coastal sharks (a group encompassing many species) per fishing trip per day. Sharks caught in Louisiana state waters are supposed to count toward the federal total quota, a science-based limit on how many fish can be sustainably caught in a year, though conservationists point out that Louisiana doesn't have the best track record of playing by the rules. The state of Louisiana also closes shark fishing in state waters during the months of April, May, and June.

States may also have a list of prohibited shark species fishers aren't allowed to catch. Most states' lists of prohibited species are the same as those used in adjacent federal waters. This is because the government does a pretty good job at creating these lists and because it's much simpler to use an existing list, especially for fishing vessels that operate within both state and federal waters. Additionally, state officials serve on federal advisory panels, so lots of people have a vested interest in having consistent rules throughout a species' range. One exception where this is not the case is in Florida, where species including lemon sharks, tiger sharks, and three species of hammerheads are protected in state waters but *not* in adjacent federal waters.

An analysis performed by my former labmate Fiona Graham found that great hammerhead sharks—which are protected in Florida waters but not in adjacent federal waters—spend about 18% of their time in Florida state waters. Tiger sharks—also protected in Florida waters but not in adjacent federal waters—are in better conservation shape than hammerheads, but only spend about 2.3% of their time in Florida state waters. This suggests that the benefit of being protected only in Florida state waters may be marginal if the shark roams too far. Let's also consider the case of lemon sharks, which famously cluster off the coast of Jupiter, Florida, every year. Their aggregation straddles the line between state and federal waters, so being protected in Florida waters only is perhaps not especially helpful if the seasonal gathering swims back and forth between the two sides of the virtual demarcation. A weird side effect of Florida's stronger-than-federal list of prohibited species is that anglers can catch species prohibited by the state if they just sail over into the federal waters that begin just outside of Florida waters. They can also still land these species in Florida ports as long as their boat doesn't stop on the way back in.

In addition to quotas, size limits, and prohibited species lists, there are also state-level laws that can affect sharks, such as shark fin trade bans. A side note from my life: the relative difficulty of obtaining a research permit to study sharks varies pretty widely between states as well.

Interstate Regulations and Fisheries Management Councils

Many fish stocks span the waters of more than one state, or cross between state and federal waters. (Most fish probably don't even know that our political boundaries exist—but I've always suspected that striped bass understand a lot more about human society than they're letting on.) Rather than just having a chaotic patchwork system where the rules are different if you're on one side of an arbitrary line, the United States has established eight Fishery Management Councils to "allow regional participatory governance by knowledgeable people with a stake in fishery management." In addition to the boundaries shown on the map below, we also have the Caribbean Council (Puerto Rico and the US Virgin Islands), the North Pacific Council (Alaska, Washington, and Oregon), and the Western Pacific Council (which includes Hawaii as well as Guam, the Northern Mariana Islands, American Samoa, and some smaller islands like Midway). Some states participate in two councils. North Carolina, for instance, falls into both the Mid-Atlantic and South Atlantic Councils, while Florida is in both the South Atlantic and Gulf Councils.

Each council creates and updates fishery management plans that govern fishing in their region's particular federal waters, though some council-created fisheries management plans are jointly written with state agencies and interstate commissions to ensure complementary regulations. To accomplish this, each council is made up of representatives from the fisheries and wildlife management agency of each of the included states, as well as representatives from the National Oceanic and Atmospheric Administration. NOAA (pronounced like "Noah" and never referred to as "the NOAA," despite what you may have heard on *The West Wing*) is the US federal agency that manages fisheries. Additionally, council members include representatives from the fishing industry and the environmental nonprofit community, as well as independent academics. For example, the Mid-Atlantic Fishery Management Council has 21 voting members, including one representative responsible for marine fisheries issues for the states of New York, New Jersey, Pennsylvania, Delaware, Virginia, Maryland, and North Carolina. It also includes a representative from NOAA's Greater Atlantic Regional Fisheries office and 13 appointed members who represent industry or the environmental community. The councils have non-voting members who can participate in discussions, as well as support staff and expert advisory panels. Each council also has a committee called a Scientific and Statistical Committee to help ensure that fishing limits are in line with federal standards. These committees can also conduct research to facilitate greater understanding of the fish stocks they manage and the effects of proposed regulations. Their committee reports are all publicly available, though I'll warn you they can get quite technical.

The United States's largest shark fishery, the Atlantic Spiny Dogfish Fishery, is managed by the Mid-Atlantic Council in cooperation with the New England Council. Together, these councils impose a combination of commercial fishery trip limits (how many sharks a boat can keep at a time) and total allowable catch (how many total sharks can be caught by the whole fishery in a year). In 2020, the quota was 23.2 million pounds of dogfish; in 2021, this number is set to increase to between 27 and 30 million pounds, though for this specific fishery, that much is unlikely to be caught due to limits on demand.

I sometimes am asked, "Are the councils good?" Well, that's a matter of debate. Many scientists and environmentalists have expressed concerns about their decisions and the decision-making process itself, which perhaps stems from the fact that industry typically has more representation than environmental groups on the councils. That said, this model is definitely more public and participatory than those used by many nations. All council meetings are open to the public. This allows ordinary citizens, scientists, or environmental campaigners to speak in support of or opposition to proposed fishery regulation changes.* Lots of important decisions happen at this level, and public support (or opposition) can make a big difference. When there's an important discussion coming up, I always post about it on social media so folks know how to help.

---

*The same is usually true of state and commission level fisheries management in the United States.

National-Level Fisheries Management Agencies

The Interior Department is in charge of salmon while they're in freshwater, but the Commerce Department handles them when they're in saltwater. I hear it gets even more complicated once they're smoked.

—President Barack Obama in his 2011 State of the Union Speech

The National Marine Fisheries Service (NMFS, pronounced "nymphs"), which is part of NOAA, is the US government office that handles commercial fisheries in federal waters. NMFS has many roles in US fisheries, including representing the councils, creating management plans for fisheries entirely within federal waters, protecting endangered species, negotiating international agreements, and issuing permits to fishers (directed shark permits for fishers targeting sharks and incidental permits for those not targeting but still catching and selling sharks). NMFS also conducts science, funds scientific research done by external academic partners, and is responsible for law enforcement. The service's Atlantic Highly Migratory Species division (HMS) deals with shark fisheries, both recreational and commercial, in federal waters (other than spiny dogfish). There's also a Pacific HMS division, but I've never interacted with them in my work.

Many shark species are caught in similar fisheries and have similar management issues, so they are grouped together into two "complexes" for convenience. The large coastal shark (LCS) complex includes silky sharks, tiger sharks, blacktip sharks, bull sharks, and spinner, lemon, and nurse sharks, as well as three species of hammerhead (great, smooth, and scalloped). Because we're often more concerned about the population status of hammerheads than some of the other species on this list, and because these sharks are more fragile and often die when caught, preventing safe release, the quota for the whole complex is linked to theirs. Oh, and blacktip sharks are only part of the LCS complex in the Atlantic; in the Gulf of Mexico, they're managed separately. (This is complex and technical, but it matters, I promise. For reading so diligently, here's a fun shark fact for you: one time a lemon shark swallowed a sharp piece of metal, and over the course of several weeks pushed the shard out the side of its body, which promptly healed, leaving no trace of a wound. There are links to a story with photos of this on the book's website.)

The small coastal shark (SCS) complex includes the Atlantic sharp-nose, bonnethead, blacknose, and finetooth sharks. In my experience with public science engagement, many shark conservation enthusiasts haven't even heard of many of these species, which is probably because they're in less conservation concern than other larger species. That's a shame both because their management issues matter and because some of them are awesome. Bonnetheads, for instance, are the only species of shark capable of digesting plants. Atlantic sharpnose sharks are my second-favorite species of shark because they're really small but they go for bait bigger than their whole body—there's something to admire about that attitude. The small coastal complex's quota is linked to that of blacknose sharks because that is the species whose populations we're most concerned about in this group at the moment. Other sharks managed by NOAA's Highly Migratory Species branch include smooth-hounds, blue sharks, porbeagle sharks, and a group called "pelagic sharks other than porbeagle or blue sharks" (which includes thresher sharks, shortfin mako sharks, and oceanic whitetip sharks). The branch's responsibilities also include managing the shark research fishery.

The HMS fishery holds an advisory panel meeting twice a year. Like the councils, it includes fishers, academics, and environmentalists. (There are far more seats for environmentalists and academics at this meeting than within the councils.) It's also open to the public and takes place three blocks from where I'm writing this. Though it's an organized and formal event, there are some nice personal touches, including the fact that one fisherman's wife regularly sends him with enough home-made desserts for everyone (which is typically well over 100 people). Hanging out at the hotel bar after the 2019 HMS advisory panel meeting was one of the most fascinating experiences of my career. After spending all day yelling at each other, everyone was not only cordial, but super friendly. There's a lot of performative outrage in stakeholder meetings— people who like each other and know we're all basically on the same side posture at being furious at each other to make a point. I had long heard about this cultural quirk, but had never seen it illustrated quite so starkly. (Of course, there are definitely people who genuinely do not like each other in the group, but the overall dynamic is not as extreme as it appears during the negotiations!)

NOAA creates the national-level prohibited shark species list. It is the agency that deals with endangered species in the ocean—more on that in the next chapter. But national legislatures can also pass laws that affect sharks without involving national fishery management agencies, though Congress often checks with NOAA.

The Shark Research Fishery, for example, is a unique partnership between government, industry, and academia. Since 2008, a small number of fishers have been granted permits to catch otherwise protected species (mostly but not exclusively, the sandbar shark #BestShark) as long as they collect data and scientific samples for researchers who are studying these species. This also allows us to explore how different fishing gear or methods affect sharks. Having 100% observer coverage produces insights that may improve management elsewhere. Some shark fishers call the Shark Research Fishery the "guinea pig fishery" because of this opportunity for experimentation, which provides really important data that supports stock assessments for several species. In a 2019 NOAA Fisheries blog post about the fishery, participating fisherman Charlie Locke was quoted as saying that "It's a place where we can find more effective practices and continue the United States' long history of being a model of sustainable shark management for the rest of the world." (If you want to know what commercial shark fishing looks like in the United States, I encourage you to check out Charlie's Instagram page, @FVsalvation. His photos are sometimes a little graphic, but he has a fascinating perspective.) To learn more about this fishery, you can also read my January 2020 essay for *Scientific American*, "To Save Endangered Sharks, You Sometimes Need to Kill a Few."

International Scale: Regional Fisheries Management Organizations

> How inappropriate to call this planet Earth when it is quite clearly Ocean.
>
> —Arthur C. Clarke

The more political jurisdictions get involved, the more complicated fisheries management gets. But what happens when a fish doesn't even stay within the waters of a whole country, the 200 nautical mile EEZ (exclusive economic zone) we talked about earlier? What happens when a fish moves between the EEZs of two countries, between, say, New England waters and Canadian waters, or between the waters of Texas and Mexico? Or even worse, what happens when a fish enters the dreaded high seas, international waters, or as we largely call them today, "areas beyond national jurisdiction"? Is the open ocean a lawless wasteland with no rules, a place where anyone can do anything?

Well, no. Certainly boundary-spanning stocks make things a little more complicated, but there are many rules governing what happens in these cases. (Which is good, because as Arthur C. Clarke noted, most of the surface of the Earth is ocean.) In the fisheries world, what happens here is mostly governed by Regional Fisheries Management Organizations (RFMOs), which are treaty-established bodies that manage fisheries whose populations span multiple political boundaries or include areas beyond national jurisdictions. RFMO member nations usually include countries whose territorial waters straddle a fish stock's range, or those whose fishers target those fish in areas beyond national jurisdiction, which are called *distant water fishing nations*. There are no shark-focused RFMOs, but four major tuna RFMOs (see Plate 10 in the color insert) deal with sharks, mostly as a bycatch issue. Several RFMOs have also set shark landings limits or bans or restrict shark finning.

The four tuna RFMOs that deal with sharks are ICCAT (the International Commission for the Conservation of Atlantic Tunas), IATTC (the Inter-American Tropical Tuna Commission), WCPFC (the Western and Central Pacific Fisheries Commission), and IOTC (the Indian

Ocean Tuna Commission). In addition to setting management rules that member nations agree to abide by, these RFMOs have scientific committees that evaluate existing data, assess population status, and issue nonbinding recommendations. The simplest example of an RFMO regulation dealing with sharks is a 2011 IATTC binding resolution on the conservation of oceanic whitetip sharks, a species caught as bycatch in the region. This resolution, which was implemented to deal with the problem of whitetip sharks being caught as bycatch, states that IATTC member nations' fishers "shall prohibit retaining onboard, transhipping, landing, storing, selling, or offering for sale any part or whole carcass of oceanic whitetip sharks in the fisheries . . . [and] promptly release unharmed, to the extent practicable, whitetip sharks when brought alongside the vessel." Other RFMOs have similar protections for oceanic whitetip sharks. Interestingly, these shark bycatch issues can dominate public perception of an issue. I've researched social media conversations surrounding several tuna RFMO meetings, and the word *mako* was used about three times as often as the word *tuna*.

Exactly how complicated is international scale shark management? In a paper on challenges and priorities in shark and ray conservation, Nick Dulvy calculated something called the *conservation complexity index* of shark species. This type of basic analysis, which suggests the magnitude of the conservation crisis facing a particular animal, is performed by multiplying the number of species of said animal by the number of national jurisdictions those species are found in. For example, there are two (or arguably three) species of African elephants (taxonomy, man) scattered across 37 countries, which results in a conservation complexity index of 74 (37 jurisdictions multiplied by two species) to 111 (37 jurisdictions times three species). What about sharks and their relatives, which are each found in the waters of an average of eight countries (and a maximum of 145, if we also consider blue sharks)? The conservation complexity index is over *8,000*, many orders of magnitude higher than that of elephants. That of course doesn't mean that elephants are easy to conserve, because they're not, but it means that it's much more complicated to conserve sharks and their relatives. In fact, Dulvy's paper asserts that "the complexities revealed just through this basic analysis suggest that the conservation challenge facing sharks may be one of the greatest ever tackled."

RFMO meetings are rarely open to the public, but you can certainly apply pressure to your country's delegation via a stakeholder hearing by sending an email before an RFMO meeting. Again, when the need for this arises, I share it on social media, so stay tuned.

Now that you know what theories are behind different conservation policies and which institutions have jurisdiction at each scale, we can start to talk about the conservation policies out there.

6 >> Sustainable Fisheries for Shark Conservation: Target-Based Policies

Policy Families

Earth provides enough to satisfy every man's needs, but not every man's greed.

—Mahatma Gandhi

Humans love to categorize and classify stuff to make complex concepts easier for us to understand, and environmental policies are no exception. These can be sorted in lots of different ways, including whether they target resource supply or demand, what jurisdiction they address (local versus global), whether they incorporate a spatial component (for example, protecting a specific area), and more. The way that I most often think about these regulatory tools is by placing them within two categories I call *policy families,* based on whether they emphasize staying within reasonable targets or enforcing hard limits.

The *target-based* family of policies aims to maximize sustainable exploitation and trade without promoting unsustainable overexploitation. These policies tend to be implemented for species with relatively healthy populations and/or relatively fast growth rates, and include regulations like traditional fisheries management practices that allow some level of fisheries exploitation, but prevent overfishing. In the United States, which has some of the most sustainable shark fisheries in the world, a target-based approach is enshrined into law via the Magnuson-Stevens Act, the federal law governing fisheries management in US waters. This act requires our natural resources management agencies to promote sustainable fisheries for species whose populations can withstand being fished.

In contrast, the family of *limit-based* policies aims to ban all exploitation and trade without focusing on protecting only the most threatened populations. Establishing designated shark sanctuaries where no commercial shark fishing is allowed and banning the sale of shark fins are both limit-based conservation tools.

Although strict limit-based tools like fin trade bans and shark sanctuaries get the overwhelming majority of media attention, and seem to be the only conservation policies that the social media activists I chat with are aware of, there are *lots* of different ways that we can protect sharks. In fact, some of the target-based policies have much more scientific evidence showing that they work, as well as significantly more support from scientific experts.

Those who prefer limit-based conservation policies often do so for

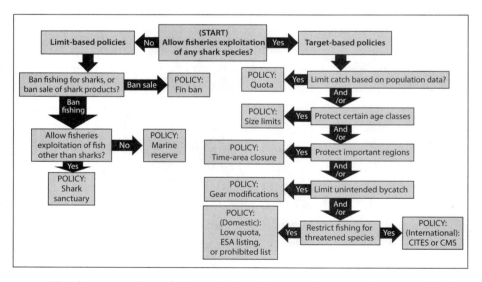

This diagram outlines which types of policies best solve particular types of shark conservation problems. Start at the top center and you can trace through to see which policies address which issues. *Courtesy of the author*

one of two reasons. Many believe, due to personal or cultural values, that no sharks should be killed for any reason, regardless of overall threats to the population or species in question. Other people, however, prefer limit-based conservation policies because they believe that sustainable shark fisheries cannot and do not exist. This is factually inaccurate. Unsustainable shark fisheries are a huge problem both historically and currently, but there's no scientific doubt whatsoever that sustainable shark fisheries can and do exist. That doesn't mean we *should* engage in sustainable fishing instead of enacting total bans on all fishing and trade. (For one thing, science doesn't usually answer questions that start with "should.") It's true, though, that some species can't withstand basically any fishing pressure. Some nations may not have the fisheries management infrastructure to do a good job managing shark fisheries.

That said, supporters of the family of target-based management tools tend to believe that not only do sustainable shark fisheries exist, but that a sustainable fisheries approach offers crucial benefits to sharks and people that bans can't provide. Allowing some sustainable fishing obviously protects the livelihoods of fishers. It also ensures that scientists can continue to catch sharks for the purpose of studying them. A perhaps less obvious but still deeply important issue is that banning shark fishing entirely can also inflict food scarcity on communities that depend on shark meat as an essential part of their diet. A 2016 survey I conducted of the members of the American Elasmobranch Society, the world's oldest and largest professional society focusing on sharks and rays, found that between 85% and 90% of shark scientists believe that sustainable shark fishing can and does exist and that sustainable management should be the goal instead of total bans whenever possible.

As we know all too well, unsustainable overfishing is the largest threat facing sharks, but that doesn't mean that shark fishing can't be sustainable. We simply need to be smart about how to manage shark fisheries. How do we do that? Well, for one, there are special considerations related to a species' biology, behavior, and ecology that must be taken into account when fishing sharks. For instance, we know that shark populations replenish more slowly due to their limited reproduction capacity than bony fish populations, which reproduce by spawning.

This means that sustainable fisheries catch limits for sharks must be lower than for a population of spawning fishes, As mentioned in the textbook *Sharks: Conservation, Governance, and Management,* "Conceptually, fisheries management of sharks is simple: the goal is to reduce mortality rates to a level that allows depleted populations to recover and other populations to maintain their productivity."

Unfortunately, as I've warned you, there is no silver bullet policy that will protect all sharks in all situations. For example, as you can see in the figure below, while a policy like a *minimum size limit* (which mandates that if you catch a shark below a certain size you must release it) is a critical component of a well-managed fishery, that specific policy doesn't do anything to, for example, reduce bycatch or protect important migratory pathways.

Research by the two former co-chairs of the IUCN Red List's shark specialist group found that only about 9% of current shark fishing around the world is biologically sustainable and well-managed. That small percentage is heavily concentrated in the United States and Aus-

	Permits	Quotas	Gear restrictions	Time/area closure	Fin: carcass ratio	Fins naturally attached	Species harvest ban	CITES Appendix I	CITES Appendix II	Endangered Species Act	Size limit	Fin bans	Marine reserve	Shark sanctuary
Allows for fisheries exploitation of some species	X	X	X	/	X	X	X	X	X	X	X	/		
Regulates total catch/control scale of fishery	X	X		/				/	/	/	/			
Regulates harvest of particularly threatened species		X								X	/			
Bans all harvest of particularly threatened species		/					X	X						
Bans harvest of all species of sharks				/									X	X
Bans all fishing in an area				/									X	
Reduces unintended bycatch			X	/									X	
Reduces inhumane and wasteful practice of finning					/	X							X	X
Restricts or bans the sale of shark fins												X		
Protects important life history stages/regions											X		X	X
Requires detailed scientific data	/	X	X	X			X	X	X	X	X			/

This figure examines the pros and cons of different approaches to shark conservation. *Courtesy of the author*

tralia. What does this mean? Some conclude on the basis of this finding that most fisheries aren't sustainable, so we should just ban all of them. However, I and most scientists I know interpret this information instead to mean that the science clearly shows sustainable fisheries can and do exist. We do still recognize that this small proportion of sustainable fisheries is inadequate, so we favor trying to help other countries' fisheries become more sustainable.

Science cannot tell us what we should do to solve a problem. However, it does tell us that sustainable fisheries absolutely can and do work if that's an option we want to pursue for political or socioeconomic reasons. Why do I support sustainable fisheries? Because they're a vital contributor to global food security and livelihoods for people. There are some shark species that cannot withstand fishing pressure; these need to be protected. But there are also some shark species that can be sustainably fished, and I don't have a problem with that. Sustainable fisheries regulations can even allow for a once-overfished population to recover (see Plate 11 in the color insert).

Fishing Quotas: How Much Fishing Is Too Much?

A *quota*, which limits how much fishers can *land,* or bring ashore, is a mainstream and traditional fisheries management tool. These quotas are usually expressed in terms of the *weight* of landed fish (e.g., 10,000 tons of fish) rather than the number of fish landed (e.g., 100,000 fish).* There are also *trip limits*, which establish how many fish a boat can catch every day or every time it leaves port. Far too many shark fisheries around the world don't currently have even this basic step in place. You'll learn about a campaign to fix this problem called No Limits No Future later in this book.

Setting science-based quotas requires significant research infrastructure. To know how many sharks can be sustainably exploited by fishers,

---

*At least to some extent, fishers have control over what they catch, but some species have to be released when caught, while others can be landed and sold.

managers need to know how many sharks are there, a process called a *stock assessment*. Managers also need to know how many baby sharks they have each year; they establish this using research disciplines known as age and growth, life history, and reproductive biology. Few people know that you can tell how old a shark is the same way you can tell how old a tree is: by counting the rings. (See "Meet a Scientist: Working in Nursery Areas and How to Age Sharks" in chapter 8 to learn more about a scientist, Bryan Frazier, who performs age and growth studies.) Only instead of examining a cross section of trunk, you count the rings on a shark's vertebrae.

There is no doubt that properly set science-based quotas can allow for sustainable fisheries exploitation and even recovery of once-depleted populations. A 2017 paper found that seven species of sharks off the southeastern United States are increasing in population despite the fact that ongoing fisheries exist for these species. (See "Meet a Scientist: Monitoring Shark Populations" in chapter 8, which presents Cassidy Peterson's efforts in shark recovery at a well-managed sustainable fishery, for more details.)

Forbidden Fruit: Species-Specific Restrictions on Fishing

Although sustainable fishing quotas are a great solution for some species, sometimes a species' population is so low that a ban on fishing for that species is required. This is considered a target-based policy because it draws on scientific data about the population or biology of that species and doesn't just restrict all fishing for all shark species. A ban means that fishing for this species is particularly restricted, but fishing for other species of sharks is still permitted in the same waters. The figure on page 132 reveals which species you cannot land in the US Atlantic and Gulf of Mexico.

In the United States, in addition to a list of species fishers must release if caught, we have a powerful conservation tool called the Endangered

Species Act, or ESA. Several species of sharks and their relatives are currently listed under the ESA, including Argentine, common, sawback, smoothback, and spiny angelsharks. Daggernose sharks, some populations of scalloped hammerhead sharks, striped smooth-hound sharks, and narrownose smooth-hound sharks also appear on the list. These species are also identified as in some degree of conservation trouble by IUCN Red List assessments, although again, a Red List assessment does not automatically convey ESA protections. Because these species are not found in US waters, protections center around restricting trade (well, some *distinct population segments* of scalloped hammerhead sharks are found in US waters, but not the ones listed by the ESA). A new addition to the list is the oceanic whitetip shark, though like the scalloped hammerhead, the US population of this species is not listed. This means that while the United States can try to promote recovery around the world, our conservation plans do not generally affect what happens to this species outside US waters. Some shark relatives like sawfish and guitarfish are also listed by the ESA. In fact, the smalltooth sawfish was the first chondrichthyan fish listed. Since smalltooth sawfish *are* found in US waters, there's a comprehensive Recovery Plan in place, along with strict habitat protections called *critical habitat designations*. It may be too early to tell, but these protections appear to be helping sawfish to recover.

A common source of confusion is that an ESA listing of Endangered or Threatened is *not* the same thing as an IUCN Red List assessment of Endangered or Threatened. The latter is an international scientific assessment that can but does not have to inform an ESA listing. The former is determined under US law specifically, and carries the force of law. In other words, an IUCN Red List assessment summarizes scientific data and raises the alarm, while an ESA listing confers some actual legal protections.

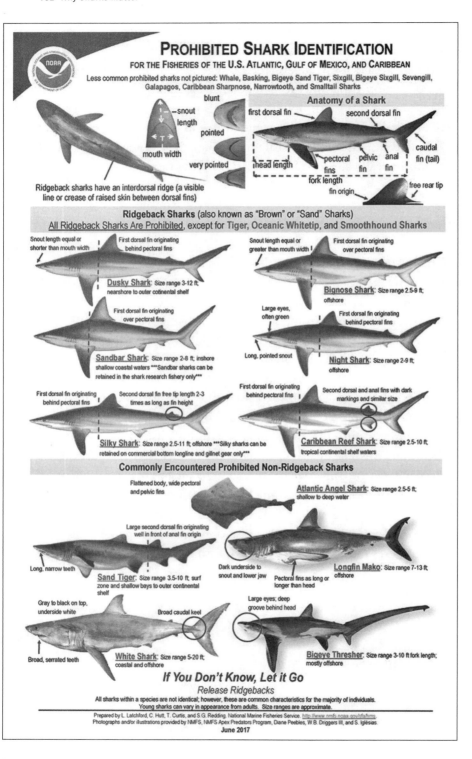

PROHIBITED SHARK IDENTIFICATION

FOR THE FISHERIES OF THE U.S. ATLANTIC, GULF OF MEXICO, AND CARIBBEAN

Less common prohibited sharks not pictured: Whale, Basking, Bigeye Sand Tiger, Sixgill, Bigeye Sixgill, Sevengill, Galapagos, Caribbean Sharpnose, Narrowtooth, and Smalltail Sharks

blunt
—snout length
pointed
mouth width
very pointed

Anatomy of a Shark
first dorsal fin — second dorsal fin
caudal fin (tail)
pectoral fins — pelvic fin — anal fin
head length
fork length
fin origin
free rear tip

Ridgeback sharks have an interdorsal ridge (a visible line or crease of raised skin between dorsal fins)

Ridgeback Sharks (also known as "Brown" or "Sand" Sharks)
All Ridgeback Sharks Are Prohibited, except for Tiger, Oceanic Whitetip, and Smoothhound Sharks

Snout length equal or shorter than mouth width

First dorsal fin originating behind pectoral fins

Dusky Shark: Size range 3-12 ft; nearshore to outer cotinental shelf

First dorsal fin originating over pectoral fins

Sandbar Shark: Size range 2-8 ft; inshore shallow coastal waters ***Sandbar sharks can be retained in the shark research fishery only***

First dorsal fin originating behind pectoral fins

Second dorsal fin free tip length 2-3 times as long as fin height

Silky Shark: Size range 2.5-11 ft; offshore ***Silky sharks can be retained on commercial bottom longline and gillnet gear only***

Snout length equal or greater than mouth width

First dorsal fin originating over pectoral fins

Bignose Shark: Size range 2.5-9 ft; offshore

Large eyes, often green

Long, pointed snout

First dorsal fin originating behind pectoral fins

Night Shark: Size range 2-9 ft; offshore

First dorsal fin originating behind pectoral fins

Second dorsal and anal fins with dark markings and similar size

Caribbean Reef Shark: Size range 2.5-10 ft; tropical continental shelf waters

Commonly Encountered Prohibited Non-Ridgeback Sharks

Flattened body, wide pectoral and pelvic fins

Atlantic Angel Shark: Size range 2.5-5 ft; shallow to deep water

Large second dorsal fin originating well in front of anal fin origin

Long, narrow teeth

Sand Tiger: Size range 3.5-10 ft; surf zone and shallow bays to outer continental shelf

Dark underside to snout and lower jaw

Pectoral fins as long or longer than head

Longfin Mako: Size range 7-13 ft; offshore

Gray to black on top, underside white

Broad caudal keel

Broad, serrated teeth

White Shark: Size range 5-20 ft; coastal and offshore

Large eyes; deep groove behind head

Bigeye Thresher: Size range 3-10 ft fork length; mostly offshore

If You Don't Know, Let it Go
Release Ridgebacks

All sharks within a species are not identical; however, these are common characteristics for the majority of individuals. Young sharks can vary in appearance from adults. Size ranges are approximate.

Prepared by L. Latchford, C. Hutt, T. Curtis, and S.G. Redding. National Marine Fisheries Service. http://www.nmfs.noaa.gov/sfa/hms. Photographs and/or illustrations provided by NMFS, NMFS Apex Predators Program, Diane Peebles, W.B. Driggers III, and S. Iglésias.
June 2017

Wrong Place, Wrong Time: Bycatch Reduction

Wherever there is fishing, there is bycatch . . . a staggering amount of marine life is hauled up with the catch, and then discarded overboard dead or dying.

—World Wildlife Fund

Bycatch is a significant threat to sharks; tens of millions of sharks are accidentally caught by fishing gear that's targeting other species of fish and shellfish every year. Luckily, there are numerous ways to reduce bycatch. Please note that I said *reduce* and not *eliminate* because, except for Critically Endangered species, the goal is typically to simply reduce bycatch below levels where it causes population-level harm.

There are dozens of peer-reviewed published papers and technical reports addressing bycatch reduction for sharks alone, and hundreds more for other groups of animals. I am just going to provide some representative examples, but these alone should be enough to convince you that bycatch is an eminently solvable problem in many cases. That it remains unsolved is due to, as with many other environmental issues, a lack of political will to implement solutions, not a lack of solutions themselves.

Successful reduction is often defined as reducing non-target catch *without* reducing target catch. If a proposed solution means that tuna fishers catch both fewer sharks and fewer tuna, that solution is extremely unlikely to be adopted. Another factor is cost. Some proposed solutions would increase the cost of fishing gear significantly, also making them extremely unlikely to be adopted. As an example, I saw a proposed bycatch reduction technology that would increase the cost per fishing hook from about 40 cents to about 15 dollars, which is especially problematic when you take into account that fact that we're talking about an industry that uses tens of thousands of hooks. Just to complicate this further, not all bycatch is unwanted; some fishermen target one species,

(Opposite page) Fishers in the United States are prohibited from catching the sharks listed in this figure. *Courtesy of the National Oceanic and Atmospheric Administration*

but can still make significant profits from selling non-target catch (usually legally—but not always). Here I'm focusing on reducing unwanted bycatch.

Historical Successes with Bycatch Reduction: Dolphin-Safe Tuna and Turtle Excluder Devices

Some of the best-known ocean conservation success stories deal with bycatch reduction, which makes it all the more strange that this important topic is so rarely discussed. Perhaps no case studies are more famous than the successes of dolphin-safe tuna and turtle excluder devices.

DOLPHIN-SAFE TUNA

> Dolphin-safe tuna is great if you're a dolphin. But what if you're a tuna? Somewhere there's a tuna flopping around going, "What about me?"
>
> —Drew Carey

Global tuna fisheries are some of the most economically valuable fisheries on Earth, employing tens of thousands of people and providing millions with food. A 2020 report by the Pew Environment Group estimated that global tuna fisheries are worth around $40 billion annually. One of the more common fishing methods used to catch tuna involves a device called a *purse seine* which is a miles-long net that's deployed around a school of tuna (see "Different Types of Fishing Gear" later in this chapter for more information on this and other fishing methods). Because deploying this net is time-consuming and difficult, fishermen don't just do it randomly and hope to catch some tuna; they wait until a school of tuna is spotted.

There are three main ways to locate a school. One of them is to look for dolphins, which eat the same food as tunas and are often found nearby. Because dolphins breathe air, they have to come to the surface, which makes finding dolphins and assuming there are tuna near them easier than finding tuna underwater. Setting a purse seine net around dolphins was a common fishing method for decades, but tragically, it meant that dolphins got caught in the nets, resulting in about 150,000 dolphin deaths a year. Though the dolphin species most commonly killed by this fishing method were not an endangered or threatened species—and the loss of 150,000 dolphins a year is not enough to endanger a whole population or species of dolphins—folks love dolphins, and this issue made a lot of people angry. It resulted in a large-scale public awareness and advocacy campaign, eventually giving rise to the dolphin-safe label you now see on cans of tuna at your local grocery store. Indeed, the new method of fishing used by tuna fishermen—who now depend on a *fish aggregating device*, or FAD—no longer involves setting purse seines around dolphins. As a result, dolphin bycatch has gone way down. Is this a success story? Well, it depends on what you measure and what you care about. To quote Todd from the cartoon *Bojack Horseman*, "Hooray . . . question mark?"

As with many stories in conservation, it's not quite that simple. Though the alternative method of fishing now used by tuna fishermen is better for dolphins, it's worse for almost everything else, including several species that (unlike dolphins) really are threatened or endangered. A 1998 paper found that FAD fishing results in more than 10 times as much bycatch of oceanic whitetip sharks, assessed as Critically Endangered by the IUCN Red List, than setting purse seines around dolphins. FAD fishing also results in twice as much sea turtle bycatch, for those of you who don't particularly care about sharks but are for some reason reading this book. The Environmental Justice Foundation did the math on this and found that for every dolphin saved, 27 sharks and rays are killed, along with thousands of other small fish.

These alternative fishing methods are also perhaps not great from a tuna sustainability perspective. By setting purse seines around dolphins, fishermen caught mostly adult tuna, whereas using FADs attracts tons of small tunas that haven't yet reproduced and may not yet be legal to catch. The use of FADs causes fishers to catch more than 20 times as much small tuna as they did when they used purse seines. Finally, it's perhaps worth noting that the dolphin-safe tuna label definitely does not mean that absolutely no dolphins were killed in catching the tuna in question. It just means that a fishing method that resulted in high dolphin bycatch in the past was not used to catch these specific tuna.

I hope this is a useful example of the trade-offs inherent in conservation and how success depends on what you're measuring. Obviously, I'm not calling for a return to now-restricted fishing methods that kill hundreds of thousands of beloved charismatic animals. The point of this story is that changing fishing methods, in this case by simply changing where you decide to set your nets, results in changes in bycatch composition. In other words, while some people believe that bycatch is an unsolvable problem, changes to how and where we fish can significantly reduce bycatch.

TURTLE EXCLUDER DEVICES

Another well-known strategy for reducing bycatch comes from the shrimp trawl fisheries of the United States. Sea turtles were caught so regularly in these trawl nets that a 1990 National Academy of Sciences study concluded that "drowning in shrimp trawls kills more sea turtles than all other human activities combined." Tens of thousands of turtles a year were caught by trawls in US waters alone during the 1970s and 1980s. The solution to this problem turned out to entail modifying fishing gear by creating what's called a *turtle excluder device* (TED) in shrimp trawls. A TED is basically a trapdoor in the net. A sea turtle is heavy enough to trigger the trapdoor and escape the net, but shrimp are far lighter and cannot get free. This solution, originally invented by a fisherman to keep cannonball jellyfish out of nets, works like a charm: it has resulted in 97% fewer sea turtles being killed and only reduced the shrimp catch by 5 to 13%. Turtle excluder devices are so effective that US fisheries have been required by law to use them since 1992. Sea turtle populations are starting to recover in the southeastern USA in no small part due to this fishing gear modification—further proof that we can absolutely do something about bycatch.

Again, this story is not quite so straightforward as it's often made out to be. Despite the fact that TEDs save sea turtles without reducing shrimp catch by very much, shrimpers in the United States have long hated them. They've filed federal court cases to change the law requiring TEDs and brought legal proceedings against the World Trade Organization. Furious shrimpers have even blockaded Gulf ports and rammed US coast guard vessels.

Part of the shrimpers' anger came from a significant misunderstanding of how severely trawlers used to harm sea turtle populations and the impacts of turtle excluder devices on their shrimp catch. Often in conservation, there are multiple sides to the story, but in this case shrimpers' claims that they weren't killing very many sea turtles were demonstrably, unequivocally false. Unfortunately, environmentalists exacerbated the situation by taking a top-down, us versus them approach, demanding new laws to change fishing practices without giving fishermen a seat at the table. In consequence, fishermen came to perceive these environmentalists as outsiders trying to shut down not only the shrimpers' livelihoods but also their whole way of life. This story offers important lessons about what not to do if we want to convince fishers to adopt changes to their methods that will reduce bycatch. Nonetheless, as with the dolphin-safe tuna story, the story of turtle excluder devices clearly shows that changes to how fishers can make a big impact on bycatch.

Now that we've seen that it's totally possible to reduce bycatch by changing how or where fishermen fish, let's apply that knowledge to the world of shark conservation. This can be done in a few basic ways: changing where we fish, changing when we fish, and changing what gear we use to fish.

Shark Bycatch Reduction

CHANGING WHERE AND WHEN WE FISH

> Most of the world is covered by water. A fisherman's job is simple:
> Pick out the best parts.

—Charles Waterman

If the goal as a fisher is to catch your target species and not sharks, the obviously ideal situation involves fishing in a place with an abundance of your target species and not very many sharks. Although open-ocean sharks and open-ocean tunas and billfishes often have pretty similar habitat requirements, there are absolutely places where one is more common than the other. They also tend to prefer different depth ranges, so even at the same geographic coordinates you can target one species but not the other just by choosing how deep you fish.

A 2008 study looked at how to reduce the bycatch of silky sharks (an IUCN Red List threatened species) in Pacific tuna fisheries. To do this, researchers examined 10 years of catch data and looked for patterns showing where silky sharks were most commonly caught and where tunas were most commonly caught. The goal was to identify parts of the ocean with lots of tunas and not a lot of silky sharks, spots where tuna fishing wouldn't result in high amounts of silky shark bycatch. They concluded that the optimal place for a "No fishing allowed here" closure zone would reduce silky shark bycatch by 33.6% (and small silky shark bycatch by over 50%) while also reducing bycatch of Critically Endangered oceanic whitetip and hammerhead sharks. Furthermore, they concluded that a closure zone would only reduce target tuna catch by 12%.

A similar recommendation based on fishing depth comes from a 2007 technical report from the Indian Ocean Tuna Commission. This report found that in the Seychelles, marlin (the target species the report focused on) were never caught when the fishing gear went deeper than 100 meters. But crucially, nearly all shark bycatch happened when the fishing gear went deeper than 150 meters. In other words, by keeping fishing gear in the top 100 meters of the water column, you can catch the same amount of marlin and yet accidentally kill far fewer sharks. Restricting your fishing depth is a straightforward, effective, and easy-to-implement solution that works for everybody—a glorious but all-too-rare win-win situation.

We can also change *when* we fish, which can mean adjusting the season or just the time of day. During seasonal migrations of gummy sharks to their pupping grounds, Australia restricts fishing gear that causes high gummy shark bycatch, although this kind of fishing is allowed other times of the year in these same locations. This is called a *time area closure*. It is distinct from a no-fishing-allowed *marine reserve* because it isn't always restricting fishing. Some shark species feed more at night. That 2007 technical report from the Indian Ocean Tuna Commission found that 75% of marlin and other target species were caught during the day, while 68% of shark bycatch occurred at night. By simply restricting fishing to daylight hours, we can significantly reduce shark bycatch while still catching lots of target fish.

How long fishing gear is in the water also matters. A 2003 study found that longer *soak times* almost always lead to more shark bycatch. This is because baited hooks attract sharks from far away, which can take a while to reach the area. After 20 hours of soak time, a fisher would accidentally catch twice as many blue and oceanic whitetip sharks and nine times as many hammerheads as with a soak time of just 5 hours.

CHANGING WHAT FISHING GEAR WE USE

It has always been my private conviction that any man who puts his intelligence up against a fish and loses has it coming.

—John Steinbeck

Sometimes the answer to reducing shark bycatch without reducing target catch can be as simple as changing what bait is used on the hooks. Blue sharks regularly dine on squid, and are more likely to be attracted to a hook baited with squid than a hook baited with fish. A 2007 report found that after the Hawaiian swordfish fishery switched from using squid bait to using fish bait, they caught 36% fewer blue sharks.

Changing what material the fishing gear is made out of can make a big difference. *Longlines* are a type of fishing gear which consists of a long rope (or line) with multiple baited hooks attached to it via detachable *leaders*. These leaders can be made out of several materials of varying strength. When leaders are made out of clear, relatively weak nylon, a shark that gets caught accidentally can bite through it and escape, which they cannot do with leaders made out of stronger stainless-steel wire. As reported in a 2007 paper, using stainless steel wire as longline leaders resulted in more than three times as much bycatch of oceanic whitetip sharks, more than twice as much thresher shark bycatch, and nearly twice as much silky shark bycatch as using nylon. Fishers can also catch more tuna on nylon because they can see a wire leader in the water but can't see the transparent nylon. Some conservation solutions carry significant costs to implement, but using nylon instead of stainless steel is a win-win for longline fishermen hoping to catch tuna without catching threatened sharks.

A few modifications to *coastal gillnets*, stationary nets that fish swim into and become entangled in, can reduce shark bycatch as well. A 2009 analysis found that using larger mesh sizes, which have bigger holes in the netting, means that smaller shark species like Atlantic sharpnoses can get through without becoming stuck. However, there's a trade-off, as larger mesh sizes also result in more bonnethead sharks getting caught. These hammerhead relatives have a head that can become entangled more easily in larger openings. More promisingly, simply increasing the tension of gillnets by using a more powerful buoy (or more buoys) to pull them upright seems to result in greatly reduced bycatch of many shark species, while still allowing the nets to catch lots of the Spanish mackerel they were targeting off North Carolina.

It turns out that the aforementioned turtle excluder devices aren't just great news for turtles. Putting a grate (not a trapdoor) at the mouth of a net can keep big creatures such as sharks out while still allowing target fish to get caught. A 2012 analysis of this technology found an 88% reduction in spiny dogfish bycatch. Not only were target hake still caught, but the hake in question were in better condition because they weren't trapped in a net with lots of freaked-out dogfish thrashing around, which tends to beat up fish pretty badly. According to a 2015 paper another helpful technique that reduces bycatch both of sharks and of Red List Critically Endangered flapper skates involves removing the heavy chain that usually drags across the seafloor in front of a trawl net.

Some of the most fascinating bycatch reduction technology involves the use of magnets. Remember how sharks can sense electromagnetic fields? Well, tunas can't, so if you put a rare-earth magnet on a fishing hook, many sharks will be repelled from that hook while tunas won't notice that anything's different. This is a situation-specific solution. Essentially, magnets seem to work really well to reduce longline bycatch of several species of sharks, but for some species, especially blue sharks, it seems to make them *more* likely to get caught. It's also an expensive fix, increasing the cost per hook several times over.

There are also bycatch reduction gear changes designed not to reduce the chance that a shark will be caught, but to increase the chance that a caught shark remains alive long enough to be safely released. Chief among these are *circle hooks*, which are designed to only get caught in the outer lip of a shark. This design means that they can't cause internal damage the way a swallowed traditional hook, also known as a J hook, can. This makes a big difference in the survival of some species, but according to a 2011 study, it weirdly seems to kill *more* night sharks than J hooks, and makes no difference in the survival of other species.

None of these solutions are perfect, but all of them help in some cases with some bycatch problems.

A QUICK NOTE ON OTHER BYCATCH

Bycatch is obviously also a big problem for ocean creatures other than sharks, from the sea turtles and dolphins we talked about earlier to seabirds and other marine mammals. There are similar gear modifications that can help with those problems. Lighting up gillnets at night can help sea turtles to see them and avoid becoming entangled. Putting streamers near hooks on a longline makes seabirds less likely to go for the bait and get hooked. Noisemakers called *pingers* alert marine mammals that an obstacle to avoid is nearby.

Bycatch is a large conservation problem, but even though lots of people think there's nothing we can do about it, it's absolutely solvable. To reiterate my main point, by changing where and when fishing occurs and what gear is used to fish, the number of sharks and other ocean creatures that accidentally get caught or wounded can be significantly reduced. None of these solutions works for every species, every type of fishing gear, or every fishery, and more research is absolutely needed to resolve the best kind of solution for every specific situation. The totality of bycatch-reducing solutions we have today wouldn't fit neatly on a bumper sticker, unlike an oversimplified conservation slogan. But that

>>DIFFERENT TYPES OF FISHING GEAR<<

Purse Seines. These are large nets that are deployed around a school of fish, often tuna, through the use of small support vessels. Then the purse strings are tightened, the bottom of the net closes, and the full net is brought to a larger vessel.

FAD Fishing. This involves the use of a FAD, or fish aggregating device, which attracts fish to a specific location so that they can be picked up later by purse seines. Any floating object works as a FAD; some are as simple as floating logs while others feature all kinds of high-tech equipment to measure what's around them and report back to a support vessel. Lots of ocean dwellers are attracted to FADs, including small fish, sharks and rays, and sea turtles, which leads to high bycatch. No one really knows why so many animals are attracted to floating objects. This behavior may be a way of seeking shelter in an open ocean environment without much of it to offer. It may also be simple curiosity. Regardless of the reason, if you put something in the ocean, many animals will swim to it and some will spend lots of time there.

Trawl Nets. These are large nets dragged behind a moving vessel. Sometimes they are dragged through the middle of the water to catch swimming animals, while other times they are weighted and dragged across the seafloor to catch bottom dwellers, which can sometimes destroy sensitive bottom habitat and make scars on the seafloor so significant that they are visible from space. The destruction these nets cause has been compared to clear-cutting a forest to catch rabbits. Some have heavy "tickler chains" at the front to stir up the seafloor, which can damage sensitive habitats.

Longlines. This gear, often used to catch tuna and billfish, is basically a series of many baited hooks, often tens of thousands, each attached to a long length of rope with buoys at both ends. The part attaching the hook to the main line (every sailor knows that once a piece of rope is brought on a boat it's suddenly transformed into a line) is called a gangion. These can free float, targeting animals in the water column, or be set with anchors and target animals on the seafloor.

Gillnets. A gillnet is a stationary net that animals swim into, resulting in them becoming entangled. It's attached to the seafloor with anchors and has buoys that hold the net upright.

doesn't mean we need to throw up our hands and give up—we can (and indeed, must) push for regulations to require these changes, and we must support attempts to innovate further.

Shark Finning Bans

Recall that shark finning refers exclusively to the act of cutting the fins of a shark off at sea and discarding the carcass at sea. If a shark's carcass is brought to shore, that shark has not technically been finned, even if the fins are later removed and sold. Lots of photos of dead sharks like the one on page 144 go viral on social media accompanied by the caption "Stop shark finning." However, the photo clearly shows shark carcasses on land, which means this is not shark finning, and stopping shark finning would not stop this.

The policy measures which have been designed to address the inhumane and wasteful practice of shark finning come in two primary forms: *a fin to carcass weight ratio* and a *fins naturally attached* policy.

Fin to carcass ratios allow fishers to remove shark fins from carcasses at sea. Fishers' argument in support of this practice include that shark fins are quite rigid and it's hard to stack entire shark carcasses, so removing the fins makes it easier to store more sharks in a cargo hold. You may think this sounds an awful lot like shark finning, but the key difference is that the carcasses aren't discarded. The fins and carcasses are stored separately on board, and both are brought to shore, or landed, with the goal that the carcasses are also sold and used (though this isn't required everywhere). The ratio is supposed to keep fishers honest by ensuring that the total weight of fins landed isn't more than a particular ratio of the total weight of carcasses landed. Usually that ratio is 5%, as in, for every 100 pounds of shark carcasses landed, you can land 5 pounds of fins.

There are a few problems with the concept of fin to carcass ratios. First and foremost, not every shark's biology is such that its fins make up 5% of its weight, which can allow for some cheating. Also, when fishers catch a shark, they often don't just put the whole body in the hold: they may gut or bleed it or cut off the head, which further messes with the carcass weight. Another troublesome phenomenon is called *high grading*, wherein fishers can swap in more valuable fins from one species and more valuable meat from another species. While they're technically landing the correct ratio of fins to carcass, the truth is that they killed a lot more sharks than are being landed. If while playing a video game

These sharks are definitely dead, and their fins may end up in the fin trade, but they were not "finned." *Courtesy of Ron Waddington, Wikimedia Commons*

you've ever dropped less-valuable loot on the ground in favor of storing more-valuable loot instead of making the trek back to town to sell what you have, you understand high grading.

A fins naturally attached policy is different from a fin to carcass ratio. Instead of landing a big pile of carcasses and a big pile of fins, then relying on enforcement personnel to compare the weights, a fins naturally attached policy requires fishermen to land sharks with their fins still attached to their carcasses. This prevents high grading or other shenanigans and makes it easier to document what was caught. It is obviously easier to determine the species of an entire dead shark, with its often highly distinguishable fins intact, than to try to identify a pile of what fishermen colloquially refer to as "logs."

The gold standard finning ban policy for a sustainable, well-managed shark fishery is fins naturally attached. In the United States, almost all shark fisheries are governed by a fins naturally attached policy. Only one American shark species is governed by a fin to carcass ratio: the Atlantic smooth dogfish. The fin to carcass ratio that the fishers who target this species employ is an alarmingly high 12%, a figure that is not based on smooth dogfish biology or really any kind of scientific evidence. This high ratio was part of a compromise with a North Carolina senator during the debate over the 2010 Shark Conservation Act, which closed some other loopholes in US shark fisheries management and conservation plans. Conservationists often use this exception to point out that dogfish, despite being a far larger part of the US shark fishery than other species, don't get the same level of protection.

It's important to stress here that shark finning bans affect *how* a shark is killed (or what you can do with the body after the shark is dead) but usually have very little impact on *how many* sharks are killed if there's a market for the meat. I don't know many serious people who claim that simply restricting finning will, by itself, significantly reduce the number of sharks killed by fisheries. Additionally, even though shark finning is banned in US waters, this ban by itself does not mean that it is illegal to buy or sell shark fins. Shark fins can be provided to the marketplace as long as the actual practice of shark finning doesn't take place. Some activists in recent years have claimed that this is an unintended loophole that must be closed. To be clear, it is not an unintended loophole. Finally, it's important to understand that the stated goal of many members of the environmental and animal welfare movement in the 1990s was to stop shark finning in order to make the fin trade a little less wasteful and a little more monitorable, capitalizing on public outcry against shark finning's waste and cruelty. The core goal of advocating to ban finning at the time was not to stop the fin trade entirely. Goals can, of course, change, but misrepresenting what the original purpose of this activism was in order to bolster current efforts is an attempt to rewrite history.

National Plans of Action

Plans are nothing, planning is everything.

—Dwight D. Eisenhower

In 1999, the UN Food and Agriculture Organization's Committee on Fisheries agreed to the set of general principles that should be factored into sustainable fisheries for sharks. This was called the International Plan of Action for Sharks, and it requested that all shark fishing nations in the UN make a National Plan of Action for Sharks (or NPOA-Sharks) by 2001. NPOA-Sharks have so far only been adopted by 13 shark fishing nations, and the actual usefulness of these national plans varies widely. Most NPOA-Sharks report on existing policies rather than setting a course for adopting new ones. While it is helpful to compile comparable information for lots of countries in one place in order to easily identify policy and practice gaps that need to be fixed, just compiling data doesn't call for or even suggest any practicable actions.

Sharks and CITES

"The Most Important Conservation Event You've Never Heard of Is About to Start"

—The headline of a *Washington Post* article I wrote in 2016

The Convention on International Trade in Endangered Species of Wild Fauna and Flora—or CITES, pronounced like *sight-ease*, not *sites*—is an agreement between world governments that deals with, you guessed it, international trade in endangered (and sometimes not yet endangered) species. By "trade" here we mean commercial sale, not, like, trading baseball cards. Many species of sharks and rays are traded internationally, meaning they are caught in one country and sold to markets in another country, so a venue like CITES is extremely important. Here's how it works.

Typically every three years, representatives from the (currently) 183 signatory nations to CITES gather for a two-week-long meeting called a *conference of parties*, or COP. Representatives from environmental groups, animal welfare groups, and many affected industries, as well as independent scientific experts, including those from the IUCN, can attend, talk to representatives, and sometimes even speak in support of or in opposition to a proposal. The voting is done by the party nations' official representatives. CITES has three appendices, or formal lists, associated with the treaty. A conservationist's goal is generally to get a species of concern listed on a CITES appendix. In shark conservation, we basically only talk about Appendix I, which essentially amounts to a ban on international commercial trade for species listed, and Appendix II, which allows trade in listed species with documentation requirements aimed to encourage sustainability. In order to legally export an Appendix II listed shark species, the government is required to show that the shark or its parts were sourced from fisheries that are both legal and sustainable. To demonstrate sustainability, a document called a *non-detriment finding* is required to prove that this particular trade will not measurably harm wild populations. How useful and reliable non-detriment findings are, again, varies quite widely. They are not required to be peer reviewed and sometimes aren't even made public. And it takes a lot of political will for one country to accuse another country of violating the rules. Even when it happens, it can take years for a formal trade review to assess those allegations.

CITES COPs are among the highest-profile international wildlife conservation meetings in the world, even though most people have never heard of them. From an analysis I've conducted on how marine issues discussed at CITES are addressed in the popular press and on social media, I can tell you that the meeting attracts significant media attention and major effort by environmental nonprofit groups. Discussions surrounding CITES also attract a ton of online engagement. For example, in 2019, 42,353 tweets were posted about CITES COP 18 by 15,500 people around the world. Compare that to the mere 2,488 tweets posted by 1,351 people about CCAMLR, the Commission for the Conservation of Antarctic Marine Living Resources, an international agreement focusing on protecting Antarctica and the ocean animals that live there.

Issues related to elephant ivory tend to overwhelm the conversation at CITES, as well as its surrounding media and social media discussions. Perhaps this is unsurprising, as elephants are well-known and beloved animals in serious need of conservation attention. In recent years, though, threats to sharks and rays have been some of the highest-profile topics of conversation (see Plate 12 in the color insert for an example of nonprofit focus on threatened shark species).

Let's consider social media first. During COP 18, 11,202 CITES-related tweets mentioned elephants compared to 3,944 tweets that focused on sharks. Sharks were mentioned four times as often as giraffes (1,143 tweets) and otters (1,083 tweets), two pretty well-known and beloved mammals. In comparison, there were a measly 393 CITES tweets about sea cucumbers, a group of invertebrates that are highly threatened due to international trade.

Traditional media coverage of CITES is likewise dominated by elephant ivory–focused articles. During COP 17 and 18, 77.4% of identified articles dealt with terrestrial species; of those, 55.4% were about elephant ivory. However, sharks and rays do now represent a significant part of the coverage. Within the 22.6% of CITES-related articles about marine species during COP 17 and 18 around 73.6% dealt with sharks and rays overall; the rest focused on specific species, including mako sharks, thresher sharks, silky sharks, and a group of guitarfish and wedgefish collectively known as rhino rays. Other marine species up for CITES listings at COP 17 and 18 didn't have a single news article primarily about them. The unfortunate sea cucumber was only mentioned at all in about 2% of articles, often as a one-sentence aside.

So far, sharks listed under CITES Appendix II include basking sharks, whale sharks, great white sharks, oceanic whitetip sharks, porbeagle sharks, great, smooth, and scalloped hammerhead sharks, silky sharks, three species of thresher shark, and shortfin mako sharks. There are no sharks in Appendix I, but some endangered rays like sawfishes are listed there. Though sharks were discussed during 1994's COP 9, the first shark species wasn't listed in a CITES appendix until after 2000's COP 11. The last three COPs have featured shark and ray discussions, and almost all of these recent proposals have been accepted (sorry, *Potamotrygon* freshwater stingrays). As of this writing, I have no idea if there will be any shark and ray proposals discussed at COP 19 in 2022, but I wouldn't be surprised if they do appear on the agenda. If so, you'll be able to follow along live, as CITES is the only global wildlife management event that includes a free online livestream of discussions. Like all the cool kids, I stay up all night watching foreign dignitaries argue about shark conservation whenever I can.

It's important to know what CITES protections can and can't do (not to mention what they aren't even supposed to do). International trade in an Appendix II listed species requires a valid non-detriment finding. *That does not mean that killing an Appendix II listed species is illegal, and it does not mean that selling meat or fins from an Appendix II listed species is automatically illegal.* You can sell products from an Appendix II listed species domestically unless there's also a domestic ban in place, and you can sell them internationally with a non-detriment finding and associated permits. It's also worth noting that any nation can choose to "take reservations" on a CITES listing for any reason at any time, which essentially means they choose to ignore the listing for as long as that reservation is in place. CITES is a powerful tool in the shark conservation toolbox, but it cannot be our only tool. Let's move on now to the better-known, but not necessarily more effective, limit-based tools.

Plate 1. Sandbar shark (*Carcharhinus plumbeus*) #BestShark in the Gulf of Mexico exhibit at the Audubon Aquarium of the Americas in New Orleans.
Courtesy of Brian Norwood, Wikimedia

Plate 2. Participating in a research workup of a large nurse shark on a semi-submerged platform in South Florida. We take a variety of measurements, take biological samples, and attach tags. *Courtesy of the author*

Plate 3. Scalloped hammerheads (*Sphyrna lewini*) off Cocos Island, Costa Rica. *Courtesy of Betty Peters, Wikimedia Commons*

Plate 4. The author wishes to remind you that since he has been sent free clothing in exchange for a photo of himself wearing it on social media, he is technically a fashion influencer. *Courtesy of the author*

Plate 5. How big are shark livers? Here's the liver from a 5-foot-long blacktip shark next to my size 11 foot. (See my article, "The Scientific Afterlife of Sharks," for more detail on what happened to this specific animal.) *Courtesy of the author*

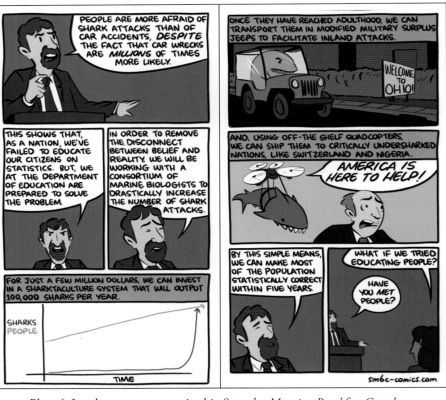

Plate 6. I make an appearance in this *Saturday Morning Breakfast Cereal* webcomic. *Courtesy of Zach Weinersmith*

Plate 7. Ottermandias says "Look upon my tummy, ye mighty, and despair!"
Courtesy of Oregon Coast Aquarium

Plate 8. (*top*) Lemon sharks in Bimini taking advantage of mangrove forests as nursery habitat. (*bottom*) Bulldozers destroying Bimini mangrove forest. *Courtesy of Kristine Stump*

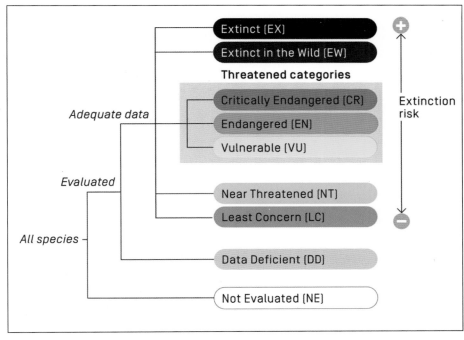

Plate 9. IUCN Red List categories. *Courtesy of the IUCN*

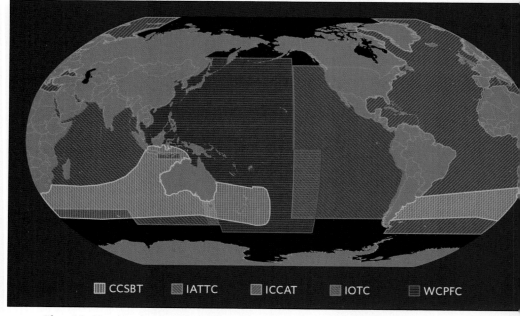

Plate 10. Graphic showing the management areas of the five tuna regional fisheries management organizations (RFMOs). *Courtesy of the Pew Environment Group*

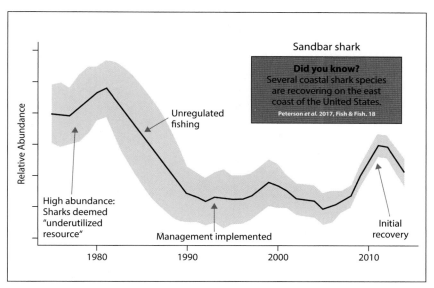

Plate 11. US shark population over time. The black line shows that, partially as a result of unregulated fishing being allowed in the United States, the number of sharks declined. Once management rules including quotas were implemented in the 1990s, the shark population slowly started to increase (but is still lower than where it started). *Courtesy of Cassidy Peterson*

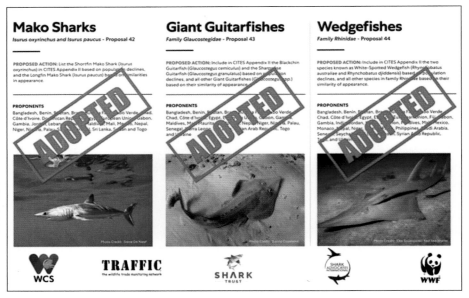

Plate 12. A social media graphic showing a successful CITES campaign to protect sharks and guitarfishes. This image was shared on social media by the Wildlife Conservation Society following the conclusion of voting. *Courtesy Shark Advocates International*

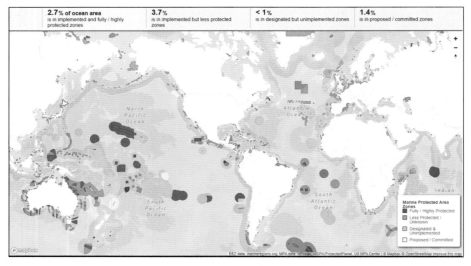

Plate 13. Sample page from the MPAtlas, a great online tool if you're interested in examining marine protection areas. *Courtesy MPAtlas*

Mako Sharks

Isurus oxyrinchus and Isurus paucus - Proposal 42

PROPOSED ACTION: List the Shortfin Mako Shark (*Isurus oxyrinchus*) in CITES Appendix II based on population declines, and the Longfin Mako Shark (*Isurus paucus*) based on similarities in appearance.

PROPONENTS

Bangladesh, Benin, Bhutan, Brazil, Burkina Faso, Cabo Verde, Chad, Côte d'Ivoire, Dominican Republic, Egypt, European Union, Gabon, Gambia, Jordan, Lebanon, Liberia, Maldives, Mali, Mexico, Nepal, Niger, Nigeria, Palau, Senegal, Sri Lanka, Sudan and Togo

Photo Credit: Steve D

Giant Guitarfishes

Family Glaucostegidae - Proposal 43

PROPOSED ACTION: Include in CITES Appendix II the Blackchin Guitarfish (*Glaucostegus cemiculus*) and the Sharpnose Guitarfish (*Glaucostegus granulatus*) based on population declines, and all other Giant Guitarfishes (*Glaucostegus* spp.) based on their similarity of appearance.

PROPONENTS

Bangladesh, Benin, Bhutan, Brazil, Burkina Faso, Cabo Verde, Chad, Côte d'Ivoire, Egypt, Ethiopia, European Union, Gabon, Gambia, Maldives, Mali, Mauritania, Mexico, Nepal, Niger, Nigeria, Palau, Senegal, Sierra Leone, Sudan, Syrian Arab Republic, Togo and Ukraine

Wedgefishes

Family Rhinidae - Proposal 44

PROPOSED ACTION: Include in CITES Appendix II the two species known as White-Spotted Wedgefish (*Rhynchobatus australiae* and *Rhynchobatus djiddensis*) based on population declines, and all other species in family Rhinidae based on their similarity of appearance.

PROPONENTS

Bangladesh, Benin, Bhutan, Brazil, Burkina Faso, Cabo Verde, Chad, Côte d'Ivoire, Egypt, Ethiopia, European Union, Fiji, Gabon, Gambia, India, Jordan, Lebanon, Maldives, Mali, Mexico, Monaco, Nepal, Niger, Nigeria, Philippines, Saudi Arabia, Senegal, Seychelles, Sudan, Syrian Arab Republic, Togo, and Ukraine

Photo Credit: Elke Bojanowski, Red Sea Sharks

Plate 12, details.

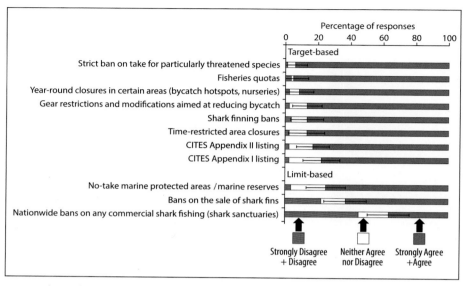

Plate 14. Respondents' level of agreement with particular shark conservation and management policies, divided into target-based and limit-based policies. *Courtesy of the Convention on International Trade in Endangered Species (CITES)*

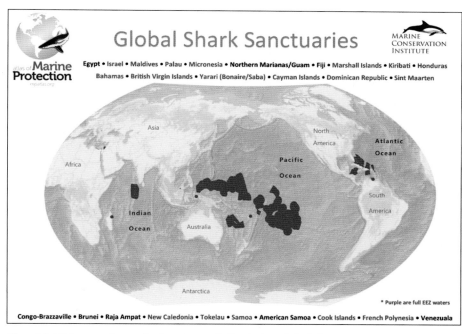

Plate 15. This map shows the locations of all existing global shark sanctuaries as of March 2018. *Courtesy of MPAtlas*

Plate 16. A Caribbean reef shark in Cuban waters investigates the bait cage in front of a baited remote underwater video station (BRUVS). *Courtesy of Global FinPrint*

7 » Fishing and Trade Bans for Shark Conservation: Limit-Based Policies

"Marine Reserves Are Necessary, But Not Sufficient, for Marine Conservation."

—Title of a 1998 paper coauthored by Jane Lubchenco, NOAA Administrator, 2009–2013

Although I suspect it's crystal clear that I tend to favor the use of target-based conservation policies for the reasons I've already given, I ultimately believe that a healthy mix of policy tools promotes healthy shark populations. Just as there's no scientific doubt that sustainable shark fisheries can and do exist, there's also no doubt that some species can't withstand basically any fishing pressure and that some countries don't have the infrastructure (or political will) to do a good job managing shark fisheries to support sustainable practices. In addition to knowing about target-based measures, you need to have a basic understanding of the second policy family—limit-based measures—so you can be an all-around thoughtful and informed advocate for sharks. To be clear, "I love sharks and I don't care if it isn't a threat to the species, I just don't want to see any shark hurt or killed ever" is a perfectly defensible perspective; it's just not one shared by me, or most of the scientific experts and environmental advocates I've interviewed in my research. It's also not the approach taken in US law, which focuses on sustainable exploitation. I often find myself perplexed by the common variant of this viewpoint: "I love sharks and I don't want to see any shark hurt or killed, but killing dogfish, skates, and rays doesn't bother me at all." This kind of differentiation seems pretty arbitrary to me.

I'm personally a lot more receptive to the argument "I'm not sure that sustainable fisheries are the right choice for this specific situation because of these reasons and this evidence" than I am to "There's no such thing as a sustainable fishery, all real scientists know that, and everyone who claims otherwise is corrupt and/or stupid and/or evil." Unfortunately, I encounter the latter line of reasoning on social media much more frequently.

Marine Protected Areas

One of the better-known limit-based measures is the establishment of *marine protected areas* (MPAs). An MPA is basically an area of the ocean where some form of human activity is restricted. You can think of them like wet national parks. These MPAs sound great for conservation, but the devil is always in the details when we're talking about complex solutions to complex global problems. The truth is that not all marine protected areas are created equal. The International Union for the Conservation of Nature (IUCN) has a set of strict definitions for protected areas on both land and sea. They sort these MPAs into categories ranging from type IA nature reserves "where human visitation, use, and impacts are strictly controlled" to type VI, where resource extraction is allowed but regulated. Sadly, some MPAs don't protect the ocean from much. An October 2020 report in the *Guardian* noted that 97% of underwater MPAs in the United Kingdom allowed bottom trawling, which makes you wonder what the point of them was in the first place. There's also a separate issue involving what are known derisively as *paper parks*, a terms which refers to areas protected on paper only that aren't meaningfully protected in practice because no one enforces the rules.

When conservation scientists and environmentalists talk about protected areas, we're usually talking about relatively strict protections closer to type IA or IB, which restricts extraction of resources but allows people to visit. Typically, when we mention marine protected areas, we're referring to a subset called *no-take marine reserves,* which means no fishing is permitted. In these areas, resource extraction, including fishing, is not allowed. You can see where they are as of this writing in a map from the Marine Protection Atlas (MPAtlas), a publicly accessible source of information on the current status of marine protected areas around the world (see Plate 13 in the color insert).

Fishing is widely acknowledged to be the greatest threat to marine biodiversity and ecosystems. Even sustainable, well-managed fisheries produce ecosystem-wide effects that can cause some harm, so we need to have some places where fishing just isn't allowed. The Bikini Atoll, for instance, where the US military tested 23 nuclear bombs over the course of 12 years, is now off limits to fishing because of radioactivity. It may shock you to hear that the atoll's marine ecosystem has recovered to the point that it has some of the healthiest coral reefs in the Pacific. To be clear, this implies that normal human fishing-related activities are worse for a coral reef than BLOWING UP A BUNCH OF NUCLEAR WEAPONS. However, because fishing provides food for billions of humans, including many of the world's poorest people, we can't (and shouldn't) just ban all fishing everywhere. Different conservation organizations vary in the specifics, but generally speaking everyone's goal is to establish a network of marine protected areas covering some of the ocean while allowing sustainable, well-managed fisheries outside of the MPAs—and restricting unsustainable fishing practices everywhere.

Properly designed, properly managed marine protected areas absolutely work to protect marine life, including sharks. A 2009 analysis found that inside the borders of MPAs, there are more fish, bigger fish, and more kinds of fish than outside the borders, where fishing is allowed. When I attended and later worked at Seacamp, we nearly always saw Caribbean reef sharks and nurse sharks when we visited Looe Key, a local MPA. We very rarely saw sharks at any nearby dive or snorkel spot that wasn't an MPA.

Effectively created and regulated MPAs also allow for overfished populations to recover and for damaged habitat to restore the ecosystem services it used to provide. The bad news is that lots of marine protected areas aren't properly designed or properly managed—but we'll get to that later. For now, the point I'd like to make is that we know that MPAs can work if we do things right, and that creating more of them is an important science-based goal. This result is not especially surprising; as my environmentalist friend Angelo Villagomez puts it, "Fish less, more fish; fish more, less fish."

A 2014 study found that the most successful MPAs have five key features. They are (1) fully no-take, (2) relatively large, (3) well-enforced, (4) relatively old (because recovery takes time), and (5) far away from humans. You also need community buy-in. A top-down approach where the government just says "You can't fish anymore" without any community consultation will result in a lot of hard feelings from fishers. In other words, if your conservation plan requires people to starve, the people you're asking to starve probably aren't going to like it. Study after study shows that people are more likely to follow environmental regulations if they feel that they have a seat at the table where those decisions are made and if they believe their concerns were listened to respectfully and truly considered. This is true even if the resulting policy outcome wasn't their first choice.

So how many MPAs do we need? The Convention on Biological Diversity (CBD), an international treaty signed by most of the world (but not the United States), included several conservation goals (called Aichi Biodiversity Targets after the city where the agreement was made) for the world to achieve by 2020. In the MPA world, we mostly talked about Target 11, which stated that "by 2020, at least 17 per cent of terrestrial and inland water, and 10 per cent of coastal and marine areas, especially areas of particular importance for biodiversity and ecosystem services, are conserved through effectively and equitably managed, ecologically representative and well connected systems of protected areas and other effective area-based conservation measures, and integrated into the wider landscapes and seascapes." Sadly, as of the end of 2020, not a single one of the Aichi Targets had been achieved—but that doesn't mean that no progress towards these goals had been made. Ultimately, it's important to keep in mind that the Aichi Targets were a starting point, not the end goal of global conservation. The current goal is no longer to protect

10% of coastal and marine areas by the end of 2020. Now, we're pushing for 30x30, which entails fully protecting (in the real-world sense, not the paper parks sense) 30% of the world's land and ocean by the year 2030. Although that represents a massive expansion in protected areas, this goal has been enthusiastically embraced by the Biden-Harris administration, and Secretary of the Interior Deb Haaland is a leader in this movement. As of this writing, I am cautiously optimistic about our chances for success in the United States.

Though most of the conservation discourse focuses just on the area targets (protecting 30% of the ocean), there's a lot of nuance in these MPA goals. For example, just banning fishing in one huge part of the open ocean wouldn't be very effective, even if we were able to protect the full 30% of its surface. In addition to this overall percentage, we also need to consider things like *ecosystem representativity*, which involves making sure that at least some of every major ecosystem is included in MPAs. Protecting all the seagrass but not including any coral reefs is no good for biodiversity goals. *Connectivity*—making sure the migratory paths between protected areas are also protected—matters as well, especially for animals that migrate or have their offspring disperse far from where they're born (like some sharks). We also need to focus efforts on the areas where these animals spend most of their time. Although people who manage terrestrial protected area networks have found that you can often put a migratory corridor basically anywhere and the animals will adjust their migration routes, that doesn't work in the ocean. There, lots of migration, not to mention larvae dispersal, is directed by currents, which means that the networks need to be in a very specific place as determined by the laws of oceanography. Finally, although it may seem obvious, protecting areas where target species actually live is critical. It's not helpful to ban all fishing in an area where no fishing occurs anyway. There are plenty of areas fishers don't frequent, which is probably a pretty solid indication that there aren't very many fish there. As a whole, we need to protect geographically spread out places that include lots of different habitat types, each of which is important to different groups of animals. We also need to take care to make sure that these protected areas are big enough and in the right place to be useful for migration and dispersal, with the ultimate goal of protecting 30% of the whole ocean.

As of this writing, according to the MPAtlas, 2.6% of the ocean is fully or highly protected. Eagle-eyed readers will notice that this figure is not only far lower than the new Aichi Target of 30% but is also quite a bit lower than 10%. But these stats are for MPAs in general. What about shark-specific MPAs?

Shark Parks: Marine Protected Areas for Sharks

In a 2018 article for the Shark Conservation Fund's blog, marine biologist Dr. Colin Simpfendorfer identified 38 MPAs specifically designed to help conserve shark populations. These MPAs took up 6% of the entire ocean surface—an area that had more than doubled over the prior two years. But wait, how can 6% of the ocean count as shark MPAs if only 2.6% of the ocean is designated protected overall? Hooray . . . question mark?

Shark-focused MPAs are a little different from the general protected areas described above. Often, they only restrict shark fishing, allowing other types of fishing, including activities that damage shark habitat or kill shark prey—even activities that accidentally kill sharks. A no-take marine reserve helps sharks (as well as other marine life) a lot more than a sharks-only MPA does. But do sharks-only MPAs work? A 2019 report jointly produced by James Cook University and the World Wildlife Fund shared a troubling conclusion: we really don't know how effective sharks-only MPAs are because very few people have checked. It's even more worrisome that we often can't assess how well many shark MPAs are accomplishing their goals because many of them *don't have stated goals anywhere* to measure success against. (This is an unfortunately common problem in conservation.)

This report noted another major difficulty: many of the world's shark-focused MPAs don't incorporate the best available scientific data and expertise—or sometimes *any* scientific data or expertise. This results in, for example, the declaration of a large area of ocean off-limits to fishing in order to protect a particular species which does not actually live there. (This is also a surprisingly widespread problem.) The report's authors also asserted that MPAs need resources for not only the planning and community consultation phase but enforcement for years to come. Just establishing an MPA without dedicating resources to protect it is a recipe for toothless protection—almost inevitably resulting in a paper park scenario. It can be particularly challenging to create an MPA in or adjacent to Small Island Developing States, especially if the size of the shark-focused MPA is larger than the total land area of the country. Think of the Maldives, for example, which occupy about 115 square miles of land, but encompass about 350,000 square miles of territorial waters. Resources are stretched thin in some small island nations, which may have only one small boat for enforcement—or even no enforcement resources at all.

Despite the fact that less-than-ideal situations do exist, we know that shark-focused marine protected areas absolutely can help—provided that they are established in the correct places, are the right sizes, have resources in place that allow them to be managed effectively, and the decision-makers creating them consult with the community to get buy-in. And of course, we need to have stated goals so we can tell if they are being achieved.

Are Marine Protected Areas in the Right Place to Protect Threatened Sharks?

I've mentioned a few times the mistake of attempting to protect a species, but instead protecting an area where that species doesn't live. This seems like a pretty obvious misstep, one that you might assume doesn't occur too frequently. However, a 2017 analysis by Lindsay Davidson and Nick Dulvy looked at the location of every single MPA and compared it to where threatened endemic species are actually found. They came to the startling conclusion that if our goal is using MPAs to protect threatened and endangered species of sharks, most MPAs are in the wrong place. Why does this mismatch keep happening? As it turns out, while many marine protected areas are designed by thoughtful professionals incorporating the best available scientific evidence as well as local stakeholder feedback, far too many MPAs are designed without these important processes and data. Often, there isn't even much thought put into where the species of interest lives. Sometimes MPA designation involves tough choices and political compromise, resulting in agreements that don't protect every area environmentalists and scientists request. But there are also times when MPA proponents just take shortcuts.

The good news is that Davidson and Dulvy's analysis found that a relatively minor expansion of these MPAs can make a huge difference for the conservation of sharks and their relatives. Just a 3% expansion in 70 countries would help protect more than half of the range of 99 species. To make this expansion even more effective, just 12 of those 70 countries are home to more than half of all those 99 species. But as of this writing, this simple adjustment has, for the most part, not happened.

Obviously, some shark-focused MPAs work better than others. Davidson and Dulvy's analysis only focused on the big picture, which clearly is not pretty. There are, however, some bright spots out there that we can learn from and build on. For example, a 2013 paper found that smooth-hound sharks spend about 75% of their time inside a protected area's boundaries. Some of them never leave. Similarly, a 2017 analysis found that two-thirds of the grey reef sharks they tracked stayed within the MPA boundary the entire length of the study—though some of the sharks swam hundreds of miles outside of the border, and several tagged sharks were killed (legally) by fishermen during the course of the study. You can tell that a tagged shark was killed by a fisher when its satellite track starts moving in a very straight line toward a port city at a constant speed. Boats move that way; sharks do not.

How Big Should MPAs Be?

Listen, if you don't talk big game, you never get anywhere. If you don't think big, you don't get big.

—Vanilla Ice

If your goal is protecting something of cultural importance like a shipwreck or a small patch of endangered corals that don't move, the MPA doesn't need to be very big at all. But if you're protecting an active, large marine species that moves several miles every day, a tiny MPA won't cut it. Work done in 2012 by some of my former labmates on Atlantic tiger shark migration patterns found that many of these animals swim from the Bahamas to Canada to Portugal and back again every year. These researchers calculated that tiger sharks' home range is equal to the size of a billion football fields. How do you protect something like that? Banning fishing in a single bay certainly won't do the trick.

A 2020 analysis of reef shark movements overlaid against the locations of coral reef MPAs found that most MPAs are simply too small to be useful for most sharks. More than one-third of the world's coral reef MPAs are less than 5 square kilometers. Although useful for protecting reef-associated species and corals themselves, such a small MPA is no good if the goal is protecting active sharks. These authors calculated that, in order to protect half of the habitat used by reef sharks in a typical day, reef-associated MPAs would need to double in size. But increasing the size of these MPAs by more than 20 times still wouldn't protect 100% of all their daily activity space.

This concern has led to the rise of the cleverly named *Very Large Marine Protected Area* (or *large-scale marine protected area*), which is defined by the IUCN as an MPA larger than 100,000 square kilometers. For your reference, the state of Delaware is about 6,500 square kilometers, so a typical VLMPA is at least 15 Delawares in size. (Please recall that the ocean is pretty big—larger than 55,000 Delawares.) The goal here is to protect an entire vast ecosystem all at once, which can include the entire range of an active species like a shark. Establishing a VLMPA also takes a measurable chunk out of the Aichi Targets (or 30x30 conservation goals) all at once. According to some conservationists who work on this issue, the positive publicity associated with establishing the "biggest MPA in the world" can be appealing to world leaders, even if it's only true until someone else makes a bigger one somewhere else. As of this writing, according to the MPAtlas database, there are 61 VLMPAs, as well as 2,576 "large" MPAs that are all between 100 and 100,000 square kilometers in size. (In comparison, there are about 10,000 very small MPAs, each of which is less than 10 square kilometers.)

The rise of the VLMPA has certainly not been without controversy among scientists and activists. The open ocean, where many VLMPAs are located, is often called a "biological desert" compared to other habitat types. There are concerns that protecting a big area where nothing much lives is not as helpful as protecting a much smaller area known to be a biodiversity hotspot, or an area that is considered a critical habitat for a particular species that's of conservation concern. In a 2018 *New York Times* op-ed, Dr. Luiz Rocha of the California Academy of Sciences criticized the rise of VLMPAs in the open ocean, arguing that "Nearshore waters have a greater diversity of species and face more immediate threats . . . If we leave these places at risk, we're not really accomplishing the goal of protecting the seas." However, proponents of VLMPAs are not saying that we should *only* focus on VLMPAs, as eloquently argued by marine conservationist Rick MacPherson in a response to Dr. Rocha's op-ed entitled "Embracing 'Yes/Also.'" (This term will no doubt sound somewhat familiar to fans and practitioners of improv comedy, which operates on a "Yes, And" principal.) MacPherson writes that "Conservation in one place in the ocean is not the enemy of conservation in another place . . . If we are going to get to the IUCN recommended target of 30% of our oceans under strong protection by 2030, we need to ramp up protections everywhere along the MPA continuum."

Another concern about VLMPAs is that they protect areas that don't really face threats to begin with. The open ocean is certainly less heavily fished than coastal seas, both because there are more fish in waters that border the coast and because it's easier for boats to get there. Dr. Rocha has noted that the locations of some VLMPAs overlap pretty closely with locations where fishing wasn't really happening even before the area was designated as protected. In other words, while we are *technically* protecting a habitat by designating it a VLMPA, it is debatable how much conservation action has actually occurred because the area wasn't threatened in the first place. Because of the perceived lack of industry opposition, such MPAs are sometimes presented as "low-hanging fruit" or easy wins, a notion that MacPherson rebuts, writing "If there are easy wins out there, big or small, I sure would appreciate someone point-ing me in their direction." And as Angelo Villagomez, who works for the marine protection arm of the Pew Charitable Trusts, chimed in, "If making MPAs like this is supposed to be so easy, then why is it so hard?" Other proponents of VLMPAs point out that while these areas don't face significant threats *now*, they may in the future, so protecting them still has value. On the other hand, there are concerns that it's pretty dif-ficult to enforce a no-fishing zone in the middle of the ocean, although emerging technologies like satellites and drones can play a helpful role here.

Nick Dulvy also wrote an article challenging the rise of VLMPAs. In it, he explained that science-based conservation of threatened spe-cies relies on lots of carefully gathered, situation-specific scientific evi-dence and argued that complex conservation problems won't be solved by simply protecting any random large area. He also expressed concern that protecting big chunks of the ocean may lead folks who are only half paying attention to ocean conservation to believe that the problem is solved when it very much is not. I've certainly observed this phenome-non anecdotally, but it is difficult to assess empirically. (Earlier, when I said that a VLMPA is 15 Delawares big, was your reaction "Wow, that sounds huge! Ocean conservation is doing better than I thought," or "That's a very tiny percentage of the total surface area of the ocean"? If the former, that's the concern Nick is talking about here.)

Finally, no meaningful discussion of this topic can be had without pointing out that one of the first VLMPAs, the Chagos Islands VLMPA in the Indian Ocean, is marred by some serious social justice implica-tions. Decades before the British government declared the area off limits to fishing, they forced the Indigenous population, known as the Chagossians, to leave their homes in order to make it easier for the United States to build the Navy Support Facility Diego Garcia. The Indigenous people weren't forcibly expelled from their traditional homes *because* of an MPA, but the later establishment of the VLMPA there rubbed salt in some very open wounds. Far too often, Western environmentalists talk about parts of the world as being "pristine," "far-off," or "remote" without noting that people live there and have for a very long time.

I understand the concerns raised by VLMPA skeptics, but I largely agree with my colleagues, who believe that VLMPAs have an important role to play in protecting the ocean. We just have to make sure that we *also* protect coastal areas while incorporating representativity, connectivity, and stakeholder perspectives. Finally, we must make sure we enforce the protections owed to VLMPAs while providing the resources necessary for that enforcement.

MPA Conclusions

Many details around the effectiveness of marine protected areas for shark conservation remain an area of active, sometimes heated debate in the science and conservation community (follow Luiz Rocha and Angelo Villagomez on Twitter and you'll see what I'm talking about). However, I think most folks on all sides of these arguments would agree with the following principles:

1. A properly established MPA can help protect many species of sharks, as well as other marine life and their habitats. MPAs will be part of the solution (but not the whole solution) to the global shark conservation crisis. They are indeed necessary but not sufficient.

2. A properly established MPA requires scientific data, community buy-in, and significant resources.

3. Whether an MPA is effective relies on a lot of different factors, and there's no one approach that always works in every situation.

4. Many current MPAs are not properly established and probably won't do very much to help sharks.

Shark Conservation and Management Policies: Shark Sanctuaries

Our traditional stories say that sharks protect the people. Now the people will protect the sharks.

—Former President Emmanuel Mori of the Federated States of Micronesia, accompanying a 2015 designation of that nation as a shark sanctuary

When a country announces that no more commercial shark fishing is to be allowed anywhere in their territorial waters or exclusive economic zone, that country is declaring their country to be a *shark sanctuary*, a term coined by the Pew Environment Group's environmental team, who worked on this issue for a decade. Some countries, including Israel and Congo, banned all shark fishing before shark sanctuaries were A Thing, and are usually not counted among nations with shark sanctuaries despite having nearly identical policy outcomes. Technically, though some advocates not affiliated with Pew refer to them by this name, a series of smaller MPAs designed to focus on protecting sharks that don't take up a country's entire exclusive economic zone aren't really shark sanctuaries.

Shark sanctuaries can be thought of as a type of marine protected area, though one just for sharks. Indeed, the shark-focused MPAs referenced earlier in this chapter include but are not limited to shark sanctuaries. However, the Pew team that worked on this issue have reported that they consider sanctuaries to operate more like a fishing quota of zero than a subset of marine protected areas. (I warned you that policy is complicated!) Some sanctuaries do not ban recreational or artisanal fishing for sharks, so there can still be shark mortality. And by allowing other types of fishing, sanctuaries still face issues with shark bycatch.

Shark sanctuaries prohibit fishing for sharks of any species, a blanket ban which is distinct from, say, the approach taken in the United States, where it's only illegal to fish for particularly threatened species of sharks. This ban emerged as a major concern about shark sanctuaries in a 2016 survey I conducted of shark scientists' perspectives on conservation tools. In the United States, many of us have the option of choosing what's for dinner, but in many fishing communities in Small Island Developing States (SIDS), you eat what you caught that day, and if you didn't catch anything, you can't eat.

In that 2016 expert survey, shark sanctuaries were the most-opposed shark conservation policy tool among scientific experts (see Plate 14 in the color insert). These experts point out that sharks can be a major contributor to food security issues and argue that a total ban on any shark fishing is overkill for what's needed. For what it's worth, that result surprised me—while imperfect, shark sanctuaries make more sense to me than banning the sale of shark fins while allowing sharks to otherwise be killed.

Because these tools are so new, there is currently no scientific evidence that they promote shark population recovery, though it stands to reason that a properly enforced sanctuary probably will eventually accomplish this. As I wrote this section, the Maldives, which created the world's second shark sanctuary in 2010, was considering ending their ban on shark fishing. The country eventually backed down in response to pressure from environmentalists, but asserted that they'd considered ending protections because local shark populations had not increased in the decade since the sanctuary was enacted. Here, sharkophile readers may be thinking, "But wait, what about the Bahamas?" Indeed, the Bahamas was declared a shark sanctuary in 2011 and has a higher shark population than much of the rest of the Caribbean. However, it's not accurate to attribute this solely to the recent shark sanctuary designation, and it's not fair to assume that other shark sanctuaries would have this same effect. The Bahamas banned commercial longline fishing 20 years before they established the sanctuary, which means they already had a relatively high shark population when they designated this protected area. It also means that the Bahamas didn't really have much of a shark fishery by the time they banned all shark fishing, which is also true of many other countries with shark sanctuaries. For example, the British Virgin Islands reported exporting about 3 tons of shark since World War II, hardly making it a fishing hotspot. Again, if VLMPAs are deployed in areas where no fishing is happening anyway, how can they truly be said to be effective at stopping fishing? Still, I'd argue that properly enforced shark sanctuaries prevent future fisheries even if they're not alleviating any current harms.

Unfortunately, at least some of the current sanctuaries *aren't* properly enforced. In addition, the people who push for sanctuary designations don't always consult local fishermen. As a result, those fishermen may feel attacked and choose to ignore the regulations. A 2014 analysis of the Maldives' shark sanctuary led by Khadeeja Ali found major problems stemming from a lack of both consultation and monitoring. Despite a ban, the analysis mentioned that it's still common to find shark products in local markets, and local scuba divers report illegal fishing that isn't investigated. The paper concluded that "commitments are needed from all stakeholders, and without regular monitoring, the shark ban cannot be a success." It has also been reported that there's still a lot of shark fishing happening in Palau, the world's first shark sanctuary, even though that's supposed to be illegal. It's no accident, by the way, that many of the world's shark sanctuaries are in the South Pacific (see Plate 15 in the color insert). Sharks are culturally revered by many South Pacific cultures, as this section's opening quote indicates.*

Shark sanctuaries were the talk of the ocean conservation world in the early and mid-2010s, but many of these efforts seem to be dying down now. Existing sanctuaries remain in place,† but new ones aren't being declared much anymore. This is mostly because the Pew Environment Group has shifted resources away from sharks since the conclusion of CITES 2019, and few other nonprofits have the resources to engage

---

*The cultural role of sharks and rays is briefly and movingly covered by the Gramma Tala character in the Disney film *Moana*.

†In March 2021, the Maldives considered revoking their shark sanctuary, but backed down after receiving significant pushback from concerned shark lovers around the world.

in national-scale conservation. It's too early to tell how much sanctuaries have helped, but early signs suggest that at least some of them are plagued by poor planning and limited resources. It's important to stress here that we wouldn't necessarily expect these sanctuaries to have a rapid and immediate detectable impact on shark populations, though, and that some were put in place to prevent future fisheries, rather than to shut down current fisheries. This means that they may well be helping in ways we just haven't found evidence for yet. But evidence matters, and thus far there's limited evidence that these shark sanctuaries have done much to help solve global-scale conservation challenges.

Shark Fin Trade Bans

Over the past decade, there has been increased environmental activist attention focused on banning the sale of shark fins. The argument here is pretty straightforward: the shark fin trade is a major reason why lots of sharks are in trouble, so let's just stop. On its surface this sounds great, and these policies have attracted a great deal of support from concerned members of the public. Indeed, most shark conservation enthusiasts I encounter assume I'll be vocally supportive of such a policy.

In actuality, I am pretty skeptical of these bans. I've even written papers for the scientific journal *Marine Policy* explaining my skepticism. This often surprises shark conservation activists I speak with, but please hear me out. (By the way, bans are still actively being considered as I write this, which means they may be the law of the land by the time you read this chapter. Still, there's value in understanding why I and many scientific experts have concerns that these policies might not do much to help.)

The first source of my skepticism is *where* the bans are being proposed. These policies aren't about banning the *global* trade in shark fins. Nor would they ban trade in shark fins in the countries that are the biggest contributors to the trade. Instead, these policies are focused on banning the sale of shark fins in the United States. Sure, the USA is a major shark-fishing nation, but we're only responsible for about 1% of the total global export of shark fins and an even smaller percentage of imports (which includes fins from sharks caught in US waters, exported abroad for processing, and re-imported).

The point here is, I think, fairly obvious: the United States generates a tiny portion of the world's shark fin trade, and the global shark fin trade doesn't look much different in scale without the United States involved. This means that banning the country's contribution to the shark fin trade doesn't (directly, anyway) save very many sharks. Most sharks caught in US waters are sold for meat, with fins supplementing the fisher's income. In one paper, my coauthor Dr. Bob Hueter and I estimated that about 23% of landed shark value in the United States comes from fins, though this obviously varies by species. About one-quarter of total value is nothing to sneeze at, but it's also not what's driving the fishery—meat sales are. Indeed, banning the sale of fins in the United States probably doesn't save any sharks. After all, a shark fin trade ban doesn't stop fishers in the country from killing sharks, it just restricts what can be done with their bodies after they're dead.

It's not just wasteful to kill sharks and throw away a part that people otherwise want, it's actually against both United Nations sustainable fisheries principles and US federal fisheries law, specifically the Magnuson-Stevens Act. In 2013 the National Marine Fisheries Service wrote the following about shark fin trade bans that were then being proposed at the state level:

> Several states have enacted or are considering enacting statutes that address shark fins. Each statute differs in its precise details, but generally most contain a prohibition on possession, landing or sale of, or other activities involving, shark fins. . . . These statutes have the potential to undermine significantly conservation and management of federal shark fisheries. State prohibitions on possession, landing, transfer, or sale of sharks or shark fins lawfully harvested seaward of state boundaries constrain the ability of federal fishery participants to make use of those sharks for commercial and other purposes. . . . If sharks are lawfully caught in federal waters, state laws that prohibit the possession and landing of those sharks with

fins naturally attached or that prohibit the sale, transfer or possession of fins from those sharks unduly interfere with achievement of Magnuson-Stevens Act purposes and objectives.
—Federal Register 78 FR 25685, posted by NOAA May 2, 2013

This argument from NOAA led to a response from the environmental community that poisoned relationships for years. Some of my NOAA colleagues still twitch a little with anger when they talk about it. An environmental group supporting the state-level shark fin bans put the ad below (which reads "NOAA: Whose side are you on? Protect sharks, not shark finners") at the DC Metro stop next to NOAA headquarters in Silver Spring. Because of its location, NOAA employees who took public transit to work were forced to see it every day. In addition to being factually inaccurate (allowing the sale of United States–caught shark fins has nothing to do with shark finning; NOAA banned shark finning in the 1990s), it ignored decades of NOAA's domestic and global leadership on shark conservation issues. In short, it pissed a lot of people off, and in the interest of discussing ideas rather than attacking people, I've cropped the name of the organization responsible out of the photo.

This ad was placed next to NOAA HQ in Washington, DC.

I'd be remiss if I did not stress that the various state-level shark fin bans are pretty inconsistent in what they actually cover. Many of them didn't cover dogfish, which are, as you recall, the most-fished sharks in the United States by far. This is like banning smoking indoors unless you're smoking cigarettes or e-cigarettes, but still calling yourself a public health hero because you banned cigar smoking. Most supporters I've spoken with had no idea that these exceptions existed.

Additionally, although the United States is responsible for a small part of the total shark fin trade, US shark fisheries provide some of the most sustainably caught shark fins in the world. If something is a major environmental problem, banning the 1% of it that's the least harmful does not make sense to me as a strategy. There's no doubt that the world has wildly unsustainable shark fisheries and that this is a major problem, but the United States does not, and that's where these bans are being discussed most frequently. Have you seen those heartbreaking videos of orangutan habitats in Borneo being bulldozed to make palm oil? If I used environmental statistics about how rainforests are burned down to make palm oil plantations as a way of arguing against kale farming in California, you'd look at me like I had two heads. But that's basically the argument being made here. Lots of shark fisheries in other places that provide fins to the marketplace are unsustainable, therefore, the reasoning goes, let's ban the sustainable ones here. This type of ban also removes a template of a well-managed fishery from the global market-place. It is easier to say "Look, we can provide shark fins using sustainable management tools, be like us" than it is to say "You should make your fisheries more sustainable. No, I don't have an example to show you, but it works, I promise." I've also heard concerns that if the United States simply bans the fin trade domestically, the federal government may become less involved in providing international pressure to convince RFMOs to enact stronger rules.

Shark fin trade bans in the United States not only don't do very much to help sharks, they take a tool that *does* help sharks off the table. The National Marine Fisheries Service offers something called *capacity building aid* in the form of expertise, equipment, or funds that allows us to help our trading partners make their fisheries more sustainable. In this case, we're not just saying, "Look, we can provide shark fins using sustainable management tools, be like us," we're saying, "We are here to help you do it." Capacity building aid is a powerful and important tool in the international shark conservation toolbox, but here's the key: the United States is only allowed to extend this aid overseas to help produce products we buy. Capacity building aid cannot be used to increase the sustainability of products that cannot be sold in US markets.

The overwhelming focus of the shark conservation community on the shark fin trade at the expense of other equal or larger threats to sharks contributes to the public misunderstanding about shark conservation threats and solutions that I've been hollering about throughout this book (and for a decade on various social media platforms). Remember the study I did of how shark conservation is discussed in the popular press? Shark fin trade bans were mentioned in newspaper articles 2.5 times as often as *all* sustainable fisheries management tools combined, despite scientific evidence and expert support for these other approaches. And my survey of shark scientists? Well, shark fin trade bans were the second most opposed policy tool after shark sanctuaries, attracting significantly more opposition from the scientific community than sustainable fisheries management tools.

When I surveyed environmental nonprofit activists, the result was even more striking. Among respondents who believe that sustainable shark fisheries do exist, 22% supported shark fin trade bans. Among respondents who believe that sustainable shark fisheries do *not* exist, 100% supported shark fin trade bans. Recall that sustainable shark fisheries *absolutely, unequivocally* exist, which means that, in other words, people who believe a thing that is wrong about shark conservation are *much* more likely to support shark fin bans. Additionally, I discovered that 75% of environmental nonprofit respondents who never read the scientific literature supported shark fin trade bans, compared with just 32% of respondents who regularly read the literature. This implies that people who know more about the facts and evidence surrounding these issues are much less likely to support shark fin trade bans. (Obviously there are plenty of smart and knowledgeable people who do support such a ban, including trusted and respected colleagues, but these trends are striking.)

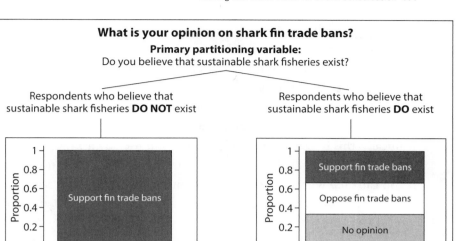

Conservationists with more understanding of scientific facts were more likely to support science-based policy solutions, as shown by this analysis of survey results from professional conservationists. *Courtesy of the author*

There's also a significant amount of implicit (or sometimes quite explicit) racism associated with discussions of banning shark fins. I've had some activists explicitly tell me that a bowl of shark fin soup is repulsive or repugnant to them, but a grilled shark steak is fine because "That's how normal people eat fish." From the perspective of a population biologist, that makes no sense; either way, a shark was killed to make food. This cultural discrimination angle has been raised in attempts to appeal state-level shark fin bans, notably in California, where the appeal was unsuccessful.

Banning the shark fin trade in the United States won't do much to directly help save sharks. It would make it harder for us to use tools that actually do save sharks while contributing to public misunderstanding about the biggest shark conservation problems and most-supported solutions. Additionally, many supporters are misinformed about the basic facts surrounding these important issues. I certainly don't mean to suggest that everyone who disagrees with me is wrong, but it is notable that almost everyone I've encountered who is demonstrably wrong about these background issues disagrees with me on this policy. But in the interest of fairness, here are some more detailed arguments for a ban and an explanation of why I object to them.

One argument for banning the fin trade is that, although some shark species can withstand sustainable fisheries exploitation, once a fin is detached from a shark you can't tell what species it came from. This raises the possibility that it might be an endangered species illegally caught. Ironically, this is essentially the same argument used by fishing groups to protest species-specific fishing restrictions, though they often insist that if you can't tell what species a shark is you can't enforce a fin trade ban. Therefore, their reasoning goes, there simply shouldn't be a ban. Both groups are wrong. You can absolutely identify restricted species both visually and genetically, though this is typically done more at the Customs level than by individual consumers. We now have handheld devices that can rapidly and cheaply identify what species a shark is from a DNA sample. These devices take little training to master (see "Meet a Scientist: Shark Fin Forensics" in chapter 8 for more information).

Another argument for banning the fin trade is that while fins can be provided to the marketplace without finning taking place, once a detached fin is for sale, it's pretty hard to tell if it came from a sustainable, well-managed fishery or was finned. Even if you can identify the species, lots of species are found in the waters of many countries, including some with sustainable fisheries management in place and others with pretty terrible fisheries management. In the abstract this argument is true, but that's why traceability standards matter. When it comes to other sustainable seafood products, we're already at the point where a packaged fillet for sale at the grocery store will tell you not only what part of the world it was caught in and with what gear but which specific boat caught it. The genetic ID techniques I mentioned earlier can identify not only what species of shark the fin came from but in what part of the world that shark was caught. Trade records and traceability standards could easily be set up so you could tell exactly where each particular fin came from. Obviously, doing this requires regulations and the political will to create and enforce them, but the notion that the identification of shark fins is an unsolvable problem is not accurate.

Fin ban proponents claim that cultural leadership can help spread support for shark fin bans. If the United States, for example, simply "leads" and says "This practice is unacceptable," other countries will follow us. Certainly, the United States has significant global cultural influence, but unfortunately, we don't have the best track record in terms of convincing other cultures that one of their cultural practices is bad and they should stop doing it. Think about Japanese whaling, which hasn't been profitable in many years and probably would have stopped on its own by now if not for the local perception that continuing it is fighting against Western imperialism.

Another argument is that we should ban shark fins in the United States because they're unhealthy. Sharks contain methyl mercury and can hold some nasty toxins. From a health perspective, you probably shouldn't be eating a lot of shark to begin with, especially if you're pregnant or may become pregnant. You've probably seen similar suggestions about tuna and swordfish made for comparable reasons. However, this is America. We don't usually ban foods on the federal level just because they're not good for you. The day I wrote this paragraph I saw a TV commercial for a new special at Kentucky Fried Chicken that comes with a bunch of donuts—this from the makers of the Double Down, a fried chicken sandwich that, instead of a traditional bun, comes nestled between two more pieces of fried chicken.* A week before that, I saw a TV commercial for pizza that uses fried mozzarella sticks for crust. A non-negligible chunk of political campaigning involves prospective leaders eating local deep-fried delicacies, including sticks of butter. The argument, then, that a particular food should be banned because it's unhealthy is frankly a little silly.

*It's real, and it's spectacular.

Shark Fin Import Bans

More and more of our imports come from overseas.

—President George W. Bush

A related policy to shark fin trade bans are shark fin *import* bans. These make it illegal to import shark fins into your country, but do not make it illegal for local fishermen to sell fins caught there or export them to other countries. This is the approach Canada currently takes. As of this writing, there's a bill in the US Congress called the Sustainable Shark Fisheries and Trade Act that aims to take this approach here instead of imposing a total ban on all US shark fin sales. Here's how this bill would work.

NOAA would conduct a thorough analysis of the shark fisheries management practices of our trading partner nations. Those that have sustainable management practices would be allowed to sell shark products (not just shark fins) in US markets. Those that do not demonstrate a commitment to sustainable practices would not be able to sell their products here. They would gain access, however, to the aforementioned capacity building aid, which would help them improve their practices so they could eventually sell their products in US markets. This is the proverbial carrot and stick approach.

I like that this bill focuses on unsustainable fisheries rather than taking a blanket ban approach that harms responsible rule-following fishermen. I also like that this bill focuses on all shark (and skate and ray) products, not just shark fins, instead of implying (or saying explicitly) that shark fins are the only problem. And I like that there's a clear path for success including actual incentives for other countries to change their practices.

However, while I see this approach as far preferable to blanket bans, there are serious concerns here. Perhaps most significantly, import restrictions open the United States up to World Trade Organization (WTO) challenges. These are expensive to fight, and there's no guarantee of victory. To avoid this, our definition of what practices count as comparable to the sustainability of US fisheries management needs to be rock solid, and it currently is not. In many ways, the United States leads the world in sustainable shark fisheries management, but we are behind in others. Recall, for example, that Atlantic smooth-hound sharks have an overly permissive fin to carcass ratio. If you tell another country "We won't buy your stuff unless you are doing as well as we are in fisheries management," then you better make sure that you are indeed doing a better job than they are in all aspects of fisheries management.

Other concerns that have been raised about the Sustainable Shark Fisheries and Trade Act include funding. NOAA's analysis of other countries' fisheries management practices would cost a lot of money, but the bill doesn't specify how it would be funded. Finally, the bill runs the risk of annoying or even alienating long-term partners on some shark conservation issues who are great at some aspects of global shark conservation but behind us in others. If we suddenly ban them from our markets, will they stop helping us on other issues that affect not only sharks but entire marine ecosystems? Additionally, NOAA already has some tools at their disposal to help put pressure on international fisheries that aren't being used to their fullest, so some of this new proposed rule may be a solution in search of a problem.

The limit-based policies described in this chapter are appealing because they present a simple, easy-to-understand solution to a problem that feels overwhelming. When it comes to some limit-based policies, though, it's important to remember that it is possible to oversimplify something so much that it is no longer useful.

8 >> How Are Scientists Helping Sharks?

Science is essential to underpin advocacy.

—An anonymous shark conservation nonprofit employee in response to a survey I distributed

P olicy is only as effective as the information it's based on, so let's discuss what kinds of information scientists provide to help policy-makers make good decisions and effectively implement them. I'll share several of the most common tools marine conservation researchers use to generate this important (and interesting) data, and I'll include some links to representative examples of publications that use those tools. I'll also introduce you to some of the brilliant and passionate people who have devoted their lives to this work. But before we explore the ways scientists are helping sharks, I want to clarify a few points of confusion about shark research that I've encountered in conversations with shark enthusiasts on social media. Even my scientific colleagues don't always seem to grasp some of these ideas.

First of all, I want to stress that not all scientific research on sharks needs to be relevant to conservation. A common query I get when I share new research from colleagues on social media is, "How does this help save sharks?" Maybe it doesn't, but that's OK; it doesn't have to. True, I am a shark conservation biologist, and this is a book about the science of shark conservation. However, there is a lot of important and valid scientific research involving sharks that has nothing to do with conservation, and that's totally fine. There are many reasons to study sharks and many questions to answer about their biology and behavior.

Researchers should not, however, claim that their science is relevant to conservation if it isn't. A corollary of the point that not all science needs to be relevant to conservation is that not all science *is* relevant to conservation. It's genuinely troubling that some scientists exaggerate the conservation relevance of their work. This is probably the most frequent thing I ding people for when peer reviewing manuscripts for scientific journals other than a failure to capitalize IUCN Red List statuses. Someone may have completed a perfectly legitimate, appropriate, and interesting scientific analysis, but then they include a throw-away line like "These results are important for shark conservation policy" with no elaboration or explanation. If a result is indeed important for conservation and management policy, a researcher should be able to answer these basic questions: What is the current regulation or policy? Why is that regulation or policy inadequate? What should the regulation or policy be instead as a result of your finding? Things also get tricky if the species being studied has been marked Data Deficient by the IUCN Red List. As you'll recall, this designation does not imply that we don't know anything at all about this species and literally any new information will save them from extinction. Instead, it means that we lack long-term population trends about the species. If a research project has revealed only random bits of information about a species facing conservation challenges, its findings will not necessarily assist in solving those challenges.

Finally, it's important to understand that solving many (though not all) conservation challenges involves finding solutions to complex technical problems. This often demands deep technical expertise and experience in obtaining and interpreting scientific data, which requires training and education. Not to put too fine a point on it, but non-experts who've been thinking about the problem for five minutes are not especially likely to come up with a better solution to a conservation challenge than scientists who've had years of rigorous training. There are absolutely ways that non-experts can help, but second-guessing experts on social media is not one of those ways, despite what are likely the best of intentions.

I hope that this chapter will give shark lovers a little more understanding of and appreciation for the detailed research that underpins modern shark conservation science. If you'd like a much more technical dive into the methods I discuss, as well as those I don't mention because they're not really relevant to conservation, check out the 2020 textbook *Shark Research: Emerging Technologies and Applications for the Field and Laboratory*, edited by Jeffrey C. Carrier, Michael R. Heithaus, and Colin Simpfendorfer.

What Is Marine Conservation Biology?

> Conservation biology, a new stage in the application of science to conservation problems, addresses the biology of species, communities, and ecosystems that are perturbed, either directly or indirectly, by human activities. . . . Its goal is to provide principles and tools for preserving biological diversity.
>
> —Michael E. Soulé, "What Is Conservation Biology?"

In the introduction to this book, I called myself a marine conservation biologist. (Depending on what audience I'm speaking to, I may also refer to myself as a shark scientist, an environmental scientist, a marine biologist, an ecologist, a fisheries scientist, or by a few other titles—wearing multiple hats is one of the interesting, if sometimes confusing, things about being an interdisciplinary researcher.) But what is marine conservation biology, and how is it different from plain old marine biology?

As initially defined in Michael Soulé's 1985 essay, "What Is Conservation Biology?" which effectively introduced my scientific discipline to the world,* conservation biology is essentially a *multi- or interdisciplinary crisis discipline*. It borrows from many related scientific fields (hence the *interdisciplinary* part) with the goal of producing applied scientific research that will help prevent species from going extinct. Because of

---

*Sadly, Michael died while I was writing this book. I'll always regret never getting the chance to meet him, though I'm a proud member of the Society for Conservation Biology, which he helped to found.

the severity of the conservation *crisis* Soulé referenced right there in his definition, it's also a discipline whose practitioners must act before all of the facts are known.

So what, then, is applied scientific research, or *applied science*? Basically, it involves answering a specific question, one with real-world relevance. It is distinct from *pure science*, which centers around learning new things for the sake of learning, whether or not they have direct benefits in the future. Often the difference between pure and applied science lies more in what the resulting data will be used for than what tools will be used to gather the data. For example, we're talking about the difference between questions like "Where does this animal go in the winter when it's not seen here?" (pure science) versus "Where does this endangered animal spend most of its time? How does that geographic range overlap with known threats? And how can we use this knowledge to design protected areas to stop that animal from going extinct?" (applied science with a goal of protecting a threatened species). Some scientists do a mix of pure and applied science. Other scientists do what I would consider conservation biology without ever calling it that. This is complicated and the lines are a little fuzzy, but if you want to learn more, I'd encourage you to check out a blog post I wrote for the Ocean Conservancy about it (you'll find the link on the book's website). Shark researchers often choose to study sharks specifically because they're endangered and need scientific research to help save them.

But there's another line here that we can't afford to leave fuzzy. It's crucial to know the difference between science in support of conservation versus advocacy. Advocacy is often considered forbidden for scientists, whose perceived neutrality is considered important if we wish to continue our vital role in the policymaking process. However, gathering the scientific data requested by policymakers to more effectively implement a policy that they have *already decided* they want to move forward on does not count as advocacy. Presenting policymakers with information that says "The data is clear that if we want Outcome 1 and not Outcome 2, we need to do Action 1, not Action 2," is also not considered advocacy (assuming that the data does clearly say that). Let's assume, of course, that the scientists in this example did not come right out and say they prefer Outcome 1. As long as decision-makers or the public have already thrown their support behind Outcome 1, scientists are perfectly free to say "If we want Option 1, Action 1 is more likely to achieve that outcome than Action 2." An example of unscientific advocacy, on the other hand, might entail scientists saying something like "Policymakers want Outcome 1, but I personally prefer Outcome 2 for reasons other than the best available scientific evidence."

It's also important to realize that advocating for the *role* of scientific data in decision-making is distinct from advocating for or against a particularly policy outcome, and is therefore not unscientific advocacy. Scientists can point out, for example, that environmental laws require the consideration of all relevant data before certain environmental policy decisions are made. Furthermore, they can point out that, in their opinion, decision-makers are not considering an important dataset while implementing environmental policy, one which would suggest a different conclusion than what a deciding body is currently leaning toward. The point here is that the importance of maintaining perceived neutrality in science certainly doesn't mean that scientists who know what's going on and how to help should never publicly speak up about anything. In fact, a growing number of younger scientists don't buy into the traditional paradigm that advocacy is unprofessional. After all, they reason, scientists are humans and care about the world just as much as other members of the community!

What Data Are Needed?

Now that I've armed you with this preliminary information, I'll discuss some of the key research needs for the conservation policies we discussed in the last chapter. I'll also introduce you to some of the research tools and methodologies used to gather data that helps answer serious conservation questions, including specific research methods used to support science-based conservation and management plans for threatened species of sharks and rays.

We need to understand which species of sharks are where at a given time and where they go when they're not "here." We also need to know how many are out there, how many babies they have, how long they live, how they interact with their environment, and how they interact with humans (and vice versa). To answer each of these questions, scientists must employ different research methods. This means that there is no one-size-fits-all conservation-relevant science. It also means that creating science-based conservation and management plans require a lot of different kinds of data gathered using a variety of methods and tools. This means that no one person (or one lab) is ever going to be a one-stop shop capable of solving the entire shark conservation crisis. Collaboration and synthesis is always needed. This is where the interdisciplinary nature of conservation science becomes critical. Conservation biology, for instance, isn't pure ecology, but uses some ecological methods and tools alongside an interdisciplinary mix of methods. Interdisciplinary, in this case, refers to a mix of natural sciences (like biology or ecology) and social sciences (like sociology, ethnography, or economics). My PhD is in interdisciplinary ecosystem science and policy. The director of my graduate program sold me on this by pointing out that to save sharks we need to study not just the sharks, but the source that threatens them the most: people.

Scientists aren't just blindly guessing what kind of data will be needed, and we're not operating in a vacuum. For example, scientists choose what research questions to address based on a variety of possible prompts. Sometimes we decide to tackle a specific question asked by a management agency that is providing funding to researchers who are willing to examine that particular issue. In these cases, the agency knows what information they need, and they're giving a scientist a grant to study that specific question and get the answer. Another scenario arises when a scientist chooses to study something based on a clearly stated public set of research priorities.* If your research question addresses an issue on a preexisting list of conservation research priorities, that's

---

*These lists of research priorities are often made by management agencies or environmental nonprofits. Occasionally, they are drawn up by academics. I generated a set of priorities for threatened sharks and rays in US waters that you can read on this book's website.

a good sign that the work you're doing is relevant to conservation and management.

If you do read a lot of the published literature in the fields I mention here, you'll quickly realize that there are probably hundreds of papers out there whose authors all did basically the exact same type of research but for a different species or in a different location. Science is community-based and iterative; we aren't lone geniuses pushing progress forward in leaps and bounds, but a large team chipping away at the boundaries of knowledge a little bit at a time.

What Species Are Found in Our Waters, and How Many Are There? Biodiversity Surveys and Stock Assessments

> Counting fish is like counting trees, except they're invisible and they move.
>
> —John Shepherd (attributed)

To understand how to effectively conserve and manage a population of sharks, we need to know what species and how many of them (approximately) there are in a given area. This may sound obvious, but in science we generally need to start with first principles—and indeed, when I lived in British Columbia, many of my neighbors had no idea that there were sharks in Canadian waters at all! In marine science, basic questions like this are often surprisingly difficult to answer.

Seriously, think about it. If I asked you to count the number of squirrels in the park by your house, you could do it relatively easily, though it'd take time. But what if I asked you to tally up the number of fish in the big lake at the center of that park? Humans just aren't built for the water, and staying underwater for more than a few moments requires expensive technology. This task becomes exponentially more logistically challenging when we're talking about going more than a few miles from shore or a few hundred feet deep. I'm always stunned when people seem surprised about the discovery of a new underwater species or the fact that we can't find a crashed plane after it was lost at sea. The ocean is really big, y'all. It makes up most of the habitat on Earth, and we can only look at a little bit of it at a time.*

One of the most basic elements of shark field research is the *population survey*, which involves using repeated consistent methods to estimate local populations and population trends. Typically, the first step in such a survey involves catching a shark, then taking measurements or samples. After that, you apply a tag that allows you to track it after you release it.

As you can likely imagine, this isn't always an easy task. You may have heard of *tonic immobility*, the tendency of sharks to briefly pass out when flipped upside down. This reflex is sometimes used by scientists and aquarium vets to quickly secure a shark for a research workup during a population survey or a medical checkup. Tonic immobility should not be used by, say, a regular person trying to get a funny photo posing with a knocked-out shark lolling upside-down (you'd be surprised at how often this happens). Although immobility can be an effective and rapid way to get sharks to hold still for trained professionals, it does cause stress to the animals. You may have heard that some people experience a peculiar sensation during a car crash or other physically traumatic event. They can see and hear everything happening around them, but can't move or speak. That's the same physiological reaction as tonic immobility. It's essentially a way of saving energy after trauma for healing or a later attempt at flight.

Some population surveys are run by the government, such as the South Carolina Department of Natural Resources (SCDNR) Coastal Shark Survey. I worked there for two years as a field assistant while completing my master's research. The SCDNR survey is a part of COASTSPAN, the Cooperative Atlantic States Shark Pupping and Nursery survey (see "Meet a Scientist: Working in Nursery Areas and How to Age Sharks" in this chapter for more information). COASTSPAN's goal is to monitor shark populations in coastal waters along the US Eastern Seaboard, with a focus on known nursery areas. Other shark population surveys are run

---

*Which doesn't mean that megalodon is still alive.

by academic institutions, like the Virginia Institute of Marine Science's (VIMS's) shark survey (see "Meet a Scientist: Monitoring Shark Populations" in this chapter for more information). Some population surveys are run by environmental nonprofits or even industry groups, though these often have a more specific geographic or species focus than those run by the government or academics.

SCDNR and VIMS population surveys are an examples of *fisheries-independent population surveys*, meaning that scientists gathered the data themselves under conditions they could control. The scientists choose which gear and bait to use, how long to leave gear in, exactly where and when to fish, etc. Interestingly, you can catch very different sharks using different gear. The SCDNR shark survey I worked with, for example, deployed research longlines and gillnets in the same spot. We would usually catch totally different species with those two different gear types, giving us a broader picture of the shark population in the area. If you're only using one type of gear, you probably won't get a complete picture of everything that's going on.

Scientists working on a population survey aren't trying to catch or see every single shark out there; that'd be just about impossible, even with a thousandfold increase in how much research is taking place. Instead, we fish the same way with the same types of gear in the same places and look for trends in what's called *CPUE,* or "catch per unit effort." If we previously caught 50 scalloped hammerhead sharks for every 1,000 hooks deployed and now catch 25 scalloped hammerhead sharks for every 1,000 hooks deployed, that's a 50% decrease in CPUE. A 50% decrease in CPUE suggests a 50% decline in the population of scalloped hammerhead sharks (or may suggest other things, including the population moving to a new habitat). If you're fishing the same way in the same place and the same number of sharks are out there, you'd probably catch just about the same number of them.

In contrast, *fisheries-dependent data* comes from analyzing data from fishing vessels' logbooks (or data from fisheries observers from onboard cameras) to look for catch trends. The advantage of this is that there are a lot more fishing vessels than there are research vessels out there, but the disadvantage is that we can't scientifically control those variables.

>>MEET A SCIENTIST<<

Monitoring Shark Populations

Dr. Cassidy Peterson, Management Strategy Evaluation Specialist,
NOAA Fisheries

Fishes are notoriously hard to monitor. Unlike terrestrial animals, we can't observe them directly because they live and move around under the water, out of view. To understand patterns in population abundances, scientists go fishing following scientifically designed protocols. We assume that the number of fishes we catch in our fishing gear is proportional to the number of fishes in the water. Essentially, if we catch more fish in a given area this year than we did last year, then we can infer that the fish population in that area is increasing. This is how we identify trends in relative population size over time.

Dr. Cassidy Peterson. *Courtesy of the subject*

Understanding trends in fish population abundance is a critical component of stock assessments, which are complex statistical models designed to provide information on the status of a fish population. Stock assessment results are used to inform the management of fish species.

The key conclusion of my work is that coastal sharks off the southeast United States are increasing, likely due to the federal management regulations on shark fishing first implemented in the early 1990s. These population increases were most notable for historically overfished large coastal species, like the sandbar shark.

I was not surprised by these results because of all the anecdotal reports of increased shark abundance. Anyone who spends time out on the water has seen the increase in the number of sharks since the 1990s. It was, however, particularly gratifying to be able to quantify this increasing trend in shark abundance and definitively state that increases in shark populations are occurring. Coastal sharks experienced large declines in abundance because of unregulated overexploitation in the 1980s. Those declines were highly (and arguably rightfully) publicized, but the severity of these declines was also heavily dramatized. This publicity, paired with the polarizing and emblematic nature of sharks, created a sort of mythos of doom and gloom surrounding sharks that percolated into popular culture. I think this deep-rooted narrative about sharks is hard to overcome now that the story is changing.

Different fishing gear modifications can reduce shark bycatch substantially, so if a fishery used to catch 20,000 sharks as bycatch per year, then changed what gear they use, and now catches 10,000 sharks as bycatch per year, does that mean that there are half as many sharks as there used to be? Or does it mean that bycatch reduction works? We have no idea; there's no way to tell simply from studying this result. This is, in fact, the basis of some of the scientific criticism of certain high-profile papers showing severe and rapid shark population declines. These articles draw on fishery-dependent data, which means we don't know if catching fewer sharks as bycatch is a result of fewer sharks being present (as the authors claim) or a result of fishers using more selective gear. A 2005 scientific paper, "Is the Collapse of Shark Populations in the Northwest Atlantic Ocean and Gulf of Mexico Real?," summarizes these concerns very thoroughly. To their credit, the authors of those original papers showing rapid and severe declines recalculated their findings based on this critique. They found that rapid and severe shark population declines are still happening, but they're not quite as bad as previously found.

Data from population surveys, as well as fishery-dependent data, goes into *stock assessments*, which are used to determine if a population or stock is overfished or is undergoing overfishing. Stock assessments tell us how many sharks are there and whether there are more or less than there used to be. You can read an example stock assessment on this book's website, but be warned, these are just about the least exciting documents associated with ocean conservation that you can imagine. One that I had to read recently opened with a seven-page-long table explaining what the acronyms used in the rest of the document meant. Sometimes it's helpful just to know whether a species is there or not and if that changes over time. This is referred to as *presence/absence data*.

Shark enthusiasts in the scuba diving community frequently ask me about the role of *citizen science* in shark population tracking. Citizen science is a kind of blanket term for supervised non-expert community participation in scientific data gathering.* Is it possible for scuba divers to help support shark science and conservation merely by monitoring and reporting what sharks they see when they go diving? When it comes to gathering presence/absence data, sure, divers can help. Just as there are more fishing vessels than there are research vessels, there are more scuba divers than there are scientists. However, many of the same cave-ats that apply to fisheries-dependent data also apply to citizen science data. Scuba divers choose where and when they go diving for reasons

---

*The term "citizen science" is falling out of favor, at least in the United States, where the concept of citizenship is increasingly loaded. After all, community members who are not citizens can certainly participate in gathering scientific data, and "community science" is an increasingly popular term for this reason. I use this term here because, as of this writing, there isn't a universally accepted alternative—and because every single project involving sharks I've seen still uses the term.

unrelated to generating consistent shark population survey data for scientists, so there are a lot of confounding variables. Not every scuba diver is equally great at identifying which shark species they saw, which means the data isn't necessarily reliable. (I for one cannot reliably tell the difference between an unusually skinny bull shark and an unusually beefy Caribbean reef shark, especially when it's far away and moving quickly.)

Another really important thing to consider is that some sharks are scared of scuba divers' bubbles, but others are attracted to them. This means that divers aren't necessarily seeing all the sharks that are present in an area; instead, they may be spotting the same individual shark multiple times as it circles around to investigate again. There are differences both between and within species in terms of reactions to scuba divers, which means there's no quick fix to this problem.

Unfortunately, the term citizen science has been corrupted by some bad actors in the non-expert shark enthusiast space. Non-scientists can absolutely contribute to scientific research, but to be effective, their work needs to be done under the supervision of, and with training from, actual scientists. Even the most enthusiastic citizen scientist, if not properly monitored, is likely to reinvent the research methods wheel (poorly) on their own. There are also for-profit businesses that charge customers money to help with supposed citizen science projects, which is, to put it mildly, not what is supposed to happen. Some research labs allow donors to participate in or observe certain projects, but citizen science is typically a volunteer operation.

One useful tool for gathering presence/absence data—and to some extent also population trend data—is a BRUVS, or *baited remote underwater video system*. Have you seen those camera traps that people put in the woods, or sometimes in their own backyard? A BRUVS is like that, but underwater. It's an underwater camera with a small crate filled with bait in front of it. Fishes are attracted to the bait and swim in front of the camera, and the footage is later analyzed in the lab (see Plate 16 in the color insert for a beautiful example of a shark swimming in front of a bait cage and getting caught on camera). These devices are typically deployed for around an hour at a time, and are most useful when you utilize a lot of them repeatedly over time (or in a lot of places). The Global

FinPrint project, for example, deployed BRUVS on hundreds of coral reefs around the world. In a recent paper, the FinPrint team analyzed over 15,000 hours of BRUVS footage and found a troubling result: there were no sharks in nearly one-fifth of coral reefs studied. These tools are also useful for studying shark behavior in places that humans can't easily access, like the deep sea or the frigid waters of the Arctic (see "Meet a Scientist: Using Cameras to Study Sharks in Inaccessible Areas" in this chapter for more information).

The newest method in the shark science toolkit is eDNA—the *e* stands for *environmental*. The premise is simple: when an organism moves through its environment, it loses some genetic material by shedding skin or excreting waste, which is detectable by scientists taking water samples. By using eDNA methods, scientists can determine if an individual of a particular species has been in the area recently—within the past couple of days, as this genetic material can degrade quickly. So far this technology has been used to supplement surveys of critically endangered, hard-to-find species. It's probably the closest thing scientists have to a *Star Trek*–style tricorder that is capable of scanning for signs of life. As amazing as eDNA methods are, this technology is more useful for performing local presence/absence studies than long-term, large-scale population trend tracking, at least for now.

Other shark research methods that use DNA include *population genetics* and *molecular ecology*. This gets technical very fast, but essentially by comparing patterns in DNA, these approaches can be used to answer many questions relevant to the management of shark species. They are commonly used for what's called *stock delineation*. A researcher might be wondering if, say, all the spiny dogfish sharks in the Atlantic Ocean constitute one giant population that should be managed together, or if there are several distinct regional dogfish shark populations that don't interbreed and need to be considered separately. DNA can help us determine whether we should consider a certain population en masse or delineate between different stocks.

>>MEET A SCIENTIST<<

Using Cameras to Study Sharks in Inaccessible Areas

Dr. Brynn Devine, Postdoctoral Fellow, University of Windsor

Working in the Arctic is an incredible privilege. From the dramatic landscapes and towering icebergs to the amazing wildlife that thrives in such an extreme environment, it's such an impressive place. I deeply enjoy the opportunity it has provided me to engage with the extraordinary people who live in northern communities and who have taught me so much about Inuit culture and the realities of living in the Arctic. The cost and logistics of getting field gear (and yourself) to remote regions in the north can be difficult, but I think the toughest thing about Arctic marine research is working around the seasonal sea ice cover. There is a limited window for sampling during the open-water season each year, and then you have the added challenge of finding ways to continue sampling through the ice in order to understand what fishes are doing throughout the year.

Baited remote underwater video systems—or BRUVS, for short—are a simple and increasingly popular method of sampling aquatic environments by photographing or recording video of species attracted to bait placed in view of a camera. As a non-extractive sampling method with little to no impact on the seafloor, BRUVS are incredibly useful for sampling sensitive habitats where traditional extractive sampling methods might be inappropriate, such as on coral reefs or within marine protected areas. BRUVS can be deployed to great depths and are less expensive than remotely operated vehicle (ROV) surveys, providing an affordable alternative for deep-sea data collection. BRUVS can be a highly efficient means of sampling; with one deployment, you can collect

Dr. Brynn Devine (above). *Courtesy of the subject*

a variety of data on both habitat characteristics and relative abundances of local fauna. The use of bait makes them most suited for sampling scavengers and predators, so these systems are also great for learning more about sharks.

I have learned that Greenland sharks love visiting BRUVS for a tasty squid treat, thus allowing us to learn more about this threatened species' summer inshore distribution and relative abundance in the Canadian Arctic. The BRUVS provided data on which depths, temperatures, and regions these sharks were most abundant in, and even identified waterways that may be particularly important for juveniles. I am excited about how much our understanding of the species has grown, along with the huge increase in public awareness in recent years. Still, I am always surprised at how many knowledge gaps remain for such a large and widespread shark species. I am excited to be working closely with industry to help tackle some of these unknowns, including how many Greenland sharks survive encounters with fishing gear and what mitigation strategies could reduce shark bycatch.

How Long Do They Live, and How Many Babies Do They Have? Life History and Reproduction

Baby shark
Doo doo, doo doo doo doo.

—Lyrics of a popular song that is much older than you think.* (Don't hate me—you know this had to go in this book somewhere.)

In order to plan population recoveries or allow for sustainable fisheries exploitation, scientists need to know where baby sharks come from. More specifically, we need to know how old sharks are when they first reproduce, how many babies they have at a time, and how often reproduction happens. Sharks, as I mentioned earlier, have relatively few

*The Pinkfong version that parents know and love is only about five years old as of this writing, but the song itself dates back over a century. I sang it at summer camp as a kid, and I don't mean Seacamp. We learned the song at J&R Jewish Day Camp in Pittsburgh in the 1990s.

offspring at a time. They give birth fairly late in life, and then only infrequently. All of this presents sustainable fisheries with management challenges.

Remember how I mentioned back in chapter 6 that you can tell how old a shark is the same way you tell how old a tree is: you cut it in half and count the rings? Annual growth bands are also found on shark vertebrae. The same is true of bony fishes, which exhibit growth bands on an ear bone that sharks don't have called an otolith. There are seasonal patterns in how much food is available to sharks, which creates a stop-and-start growth pattern, which leads to ring patterns. When lots of food is available, sharks grow more quickly. During lean times, though, they grow slowly. The older sharks get, however, the harder it is to count their growth rings; the bands start blur together, leading scientists to underestimate the age of older sharks. Unfortunately, there is no way to section vertebrae and age sharks without sacrificing them.

Determining age and growth curves that allow you to determine how old an individual of a given length is requires taking these vertebrae

This prepared lemon shark vertebrae section has 11 growth rings, indicated by the dots on the right side of its spine. The number of rings reveal that it was 11 years old when it died. *Courtesy of NOAA Southeast Fisheries Science Center*

samples, along with size measurements and such, for dozens or hundreds of individuals across a range of sizes. When trying to develop growth curves, just taking a bunch of five-foot-long blacktips and learning that they're all about 15 years old doesn't tell you much. You also need some two-foot-long blacktips, and some three-foot-long blacktips, and so on, so you can draw comparisons between individuals of different sizes and ages (see "Meet a Scientist: Working in Nursery Areas and How to Age Sharks" in this chapter for more information).

We also need to know how many babies the species being investigated has at a time, how often these sharks give birth, and how old they are when they're first reproductively mature. Technologies like blood hormone analysis and ultrasound—which relies on a waterproof, more portable version of the device used to examine pregnant humans that researchers can bring on a boat—can give some insights here. But such tools are generally considered not as useful for gathering all the data we need as examining the reproductive tract directly, at least not yet. Like studying vertebrae, examination of the reproductive tract, which gives us the most informative and accurate data, unfortunately does require sacrificing sharks. While studies of subjects like age and growth, life history, and reproduction all require bodies to study, they use different parts of the sacrificed animal. For this reason, these studies are commonly conducted simultaneously to minimize the total number of sharks we need to sacrifice while maximizing the use of each individual shark.

This work has given us answers. For example, we've discovered that about half of sandbar sharks are considered sexually mature adults (able to reproduce) by age 13 to 14. At that point, they reproduce every 2 or 3 years, and are pregnant for 9–12 months at a time. They have an average of 9 pups at a time in the Atlantic, though this number is lower in the Australian and Hawaiian populations. And the Greenland shark, which can live to be over 400 years old, doesn't reach sexual maturity until around age 150! Imagine going through puberty for as long as the modern nation of Canada has existed—that's what life is like for a Greenland shark.

>>MEET A SCIENTIST<<

Working in Nursery Areas and How to Age Sharks

Bryan Frazier, Wildlife Biologist,
South Carolina Department of Natural Resources

A shark nursery area is an area where young-of-year (newborn) sharks are found in high abundance. These areas are used every year unless unusual environmental conditions (drought, flood, or low dissolved oxygen) alter conditions beyond sharks' preferred thresholds. In South Carolina, nursery areas contain up to six species of young-of-year sharks which thrive on the abundance of prey found in nursery areas. The Cooperative Atlantic States Shark Pupping and Nursery (COASTSPAN) program was created by the National Marine Fisheries Service's Narragansett, Rhode Island, office to describe species composition and critical habitat. It also monitors abundance of young-of-year, juvenile, and adult sharks in estuarine and nearshore waters along the US East Coast.

Participating partners (including SCDNR) use longlines and gillnets to sample for sharks in these areas. Over time, these data can be used to create abundance and recruitment indices that allow us to monitor population levels. As sharks are a vital part of the coastal ecosystem and are valued by South Carolina stakeholders, recreational anglers, commercial fishers, and nature lovers, SCDNR has a keen interest in monitoring these populations as well as conducting research that allows for improved management of these important species.

Age, growth, and life history data are all critical pieces of infor-

Bryan Frazier (above). *Courtesy of Rob Nelson*

mation that can help ensure proper management of shark popula-tions. In addition to revealing basic information such as longevity (also known as maximum life span) and fecundity (how many pups are produced annually), these data can be used to model var-ious aspects of a population's demographics. We determine the age and growth by subsampling a population. Researchers sacrifice or sample commercial and recreational catch of a shark species fa-thered from throughout the population's spatial range. They also sample all sizes present in the population.

After being caught, the animals are measured and dissected. Maturity—whether the animal is juvenile or mature—is deter-mined by examining reproductive structures that are visible to the naked eye. Vertebrae are extracted, sectioned by taking a thin slice through the center, and viewed under a microscope to estimate age. Those data can then be used to model the average age and maturity at a given length for any animal in the population. These estimates are then used in stock assessments to estimate ages for all landed and released sharks for recreational, commercial, and fishery independent (scientific) surveys.

Model results (parameters) are also used to estimate other stock assessment data such as *natural mortality rates*, or the percentage of animals that are expected to survive annually independent of fisheries pressure. A shark specimen's vertebrae must be prepared prior to age estimation. Researchers manually remove muscle and connective tissue using knives and scalpels. They soak the verte-brae in bleach to remove any remaining muscle tissue. Finally, they dry the vertebrae, fix each one to a slide, and inspect the section (cut into slices about 0.4 mm thin) to estimate age.

Estimating the age of sharks is a time-consuming process. It's also smelly—residents of nearby laboratories often remark on the pungent odors associated with shark samples (shark blood contains urea, and frozen then thawed shark tissue can smell like burning pee). That said, it's something of an acquired smell. Once they get used to it, people handling samples don't really mind it and can rarely tell how far odors drift.

Where Do Sharks Spend Their Time? Telemetry Tracking

Perhaps the best-known shark science research tools are *telemetry tags*. These wearable devices broadcast the location of tagged sharks and allow scientists to, among other things, track where they go. Some newer tags also allow scientists a glimpse of what the sharks do when they get there. As this chapter demonstrates, telemetry tagging is hardly the only tool shark scientists employ, even though it is basically the only science you'll ever see mentioned on Shark Week. My conservationist friend Angelo Villagomez has joked that half of all shark science consists of a single scenario: "We put a tag on a shark and it swam pretty far away."

There are two primary types of telemetry tags: acoustic and satellite. *Satellite tags* are perhaps the best-known among shark enthusiasts due to the outreach efforts of groups like OCEARCH, as well as their frequent appearance during Shark Week. These devices are larger and more expensive than acoustic telemetry tags, and are often attached to or trailing behind the shark's highest point—the top of the dorsal fin. Satellite telemetry tags contain a variety of internal sensors that record data. When they are above the surface of the water for long enough they broadcast some of that data to orbiting satellites, which email it to researchers. Satellite tags are best known for their use in tracking migrations: each tagged shark broadcasts its GPS location wherever it goes as long as its fin is above the surface of the water for more than a few seconds at a time. These tags can also monitor depth, swimming speed, and acceleration, as well as a variety of behavioral and ecological metrics. An assortment of high-tech sensors can be miniaturized and inserted into a tag. These multisensor tags will answer a lot of interesting questions about shark behavior in the next few years. (As briefly mentioned earlier, while attaching some types of satellite tag requires drilling into a shark's fin, this is not seriously harmful if done correctly.)

Satellite tags have revealed the incredible extent of some shark migrations. A great white shark named Nicole currently holds the record for longest fish migration tracked with satellite telemetry: she swam from South Africa to Australia, a distance of about 5,000 miles, in a few months. Another example that received a lot of media attention a few years ago was a tagged basking shark that swam from New England to Brazil, the first documented case of a trans-equatorial migration for one of these giant filter feeding sharks. In recent years, OCEARCH has given individual social media accounts to some of their most famous tagged sharks, like Mary Lee and Katherine, and a satellite track showing that these sharks are nearby generates local news coverage all along the Eastern Seaboard.

Acoustic telemetry tags work a little differently. They regularly make a unique sound (or "ping"), which is picked up by an underwater receiver. These underwater receivers can only hear pings from a few hundred feet away under the best conditions, which means you need a lot of them—this is called an *array*. These arrays tend to be set up in specific locations of research interest, such as near the entrance to an estuary or around the boundary of a proposed protected area. They're quite expensive, so we can't cover the whole seafloor in arrays or deploy them at random. An individual acoustic telemetry tag is smaller than a satellite tag—it's only slightly larger than a grain of rice. It's also cheaper than a satellite tag, but can't do as much on its own: it requires a receiver array to function. Sometimes these tags are surgically inserted inside a shark's body cavity, and sometimes they're externally attached (see "Meet a Scientist: The Movement of Life" in this chapter for more information).

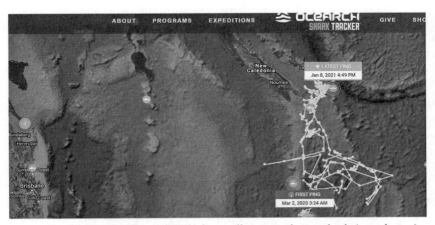

A sample track showing how David, the satellite-tagged tiger shark (no relation), moved during the course of a year. That's northeast Australia on the left. *Courtesy of OCEARCH's Shark Tracker website*

>>MEET A SCIENTIST<<

The Movement of Life

Dr. Chuck Bangley, Postdoctoral Fellow, Dalhousie University; Research Associate, Smithsonian Movement of Life Initiative

Typically, researchers deploying acoustic tags will also deploy receivers for their own studies. However, this can mean that they lose track of their tagged animals whenever they leave the study area. Receivers deployed for one particular project can also pick up tagged animals from other projects as long as their tags transmit at the same frequency. Researchers quickly realized the advantages of being able to share detection data between projects and are increasingly participating in collaborative networks such as the Atlantic Cooperative Telemetry (ACT) and Florida Atlantic Coast Telemetry (FACT) networks. Through these networks it has become possible to track marine animals along entire coastlines and even across state and international boundaries, allowing acoustic telemetry to provide data at spatial scales previously only possible using satellite tags.

It's a lot of fun seeing where the tagged animals go, especially highly migratory species like sharks. I've seen bull sharks travel pretty far up rivers, cownose rays stick around within the same estuary for an entire summer before rocketing south along the entire US east coast in the fall, and dusky sharks closely track oceanographic features I didn't even know about until I had to try to explain those movements. Getting to know the other researchers who are picking up your tagged animals is also really enjoyable

Dr. Chuck Bangley (above). *Courtesy of the subject*

and rewarding, and can lead to collaborations that teach us a lot about the movement behavior of these species and how it affects everything from population structure to feeding habits.

The movement data that we can get from acoustic telemetry alone can tell us a lot, but the ability to match that up with information on environmental conditions and human activity provides incredible potential to figure out what drives those movements and predict how our own actions might change them. Satellite-based environmental monitoring lets you look at conditions like sea surface temperature at a coast-wide scale; you can see what conditions were like both where your tagged animals appeared and, just as crucially, where they didn't. These advancements in pairing movement data with other types of data are rapidly expanding our understanding of why sharks and other marine animals go where they go.

Ideally, these tools are used to answer specific questions related to migration, habitat use and preference, and behavior. Some scientists complain that entirely too many shark telemetry studies basically involve researchers finding and tagging a shark simply to see what it would do and where it would go. This kind of approach could well lead to interesting and novel information, but isn't exactly hypothesis-driven science. On the other hand, one of the best examples I've seen of applied practical telemetry science involves using satellite-tagged sharks to help with surveillance and enforcement of large marine protected areas (see "Meet a Scientist: Making Sure Marine Protected Areas Do Their Important Job" in this chapter for more information).

>>MEET A SCIENTIST<<

Making Sure Marine Protected Areas Do Their Important Job

Dr. Meira Mizrahi, Wildlife Conservation Society

When carefully planned and managed effectively, marine protected areas can be useful tools to support marine conservation. They can also complement fisheries management. We want to have MPAs because marine ecosystems and the fisheries they support are deteriorating at an unprecedented rate. A synergy of human-generated threats—including overfishing, climate change, pollution, and species invasions—has resulted in a rapid decline in marine biodiversity, evidenced by reductions in ocean wildlife as high as 90% for some species. This decline has grave consequences for the more than 3 billion people globally who rely on marine environments for their livelihoods, more than 200 million of whom depend directly or indirectly on fisheries. MPAs are not a panacea for securing marine conservation, but they have the potential to protect important habitats and species and can aid in building the resilience and productivity of our oceans.

Unfortunately, MPAs are not always placed in areas where they can maximize their conservation impact. Several studies have shown that MPAs are generally concentrated in areas that are less economically valuable for fisheries. There is an added risk that new MPAs will be biased toward places that are remote and/or unsuitable for extractive purposes, with less consideration of the ecological or socioeconomic values of those areas. These MPAs typically

Dr. Meira Mizrahi (above). *Courtesy of the subject*

provide little protection to the ecosystems and people who are most at risk from threats to marine environments.

Social considerations are often neglected, or considered to a lesser extent, when it comes to MPA planning. We tend to think about biological and ecological factors first and foremost when designing MPAs and provide less consideration to the people who rely on those environments on a daily basis for their livelihoods and well-being. These are the people who are most severely impacted by the access restrictions of MPAs. Going forward, they should be equitably included in the MPA design process.

I'm often surprised at how universal our basic desires as human beings are to take care of ourselves and our families, put food on the table, and spend time in nature with the people we love. Most people value the natural world in which we live and have a genuine desire to ensure that biodiversity and natural resources are around for a long time. In many cases, however, the most dominant costs of conservation initiatives like MPAs are borne by local people who utilize the natural wealth of the sea for sustaining their livelihoods. For that reason, I've learned that it's so important to have empathy in conservation. Listen to people's experiences and make an effort to understand the aspirations and desires of the individuals who operate in the context within which you are working. In my opinion, MPA success is contingent on this.

It is increasingly clear that the conservation of sharks goes beyond considerations of shark and their habitats, and that really, conservation is about people. I think that shark conservation scientists and practitioners are increasingly recognizing that incorporating human dimensions into conservation planning is imperative to create conservation solutions that improve shark populations while generating tangible benefits for humanity. The critical work of shark conservation scientists will be strengthened by incorporating socioeconomic dimensions into shark MPA planning by seriously considering livelihoods, justice, equity, and inclusion.

How Many Sharks Are Being Killed/Traded? Market Surveys

The resources dedicated by a fleet of fishermen will always outmatch any scientific efforts.

—Scientist Julia Spaet, in a 2014 interview I conducted with her for *Scientific American* about the value of market surveys

The research methods described so far tell scientists how many sharks can be sustainably fished and when sharks are present in local waters. It's also important for scientists to understand how many sharks *are* being fished—and if that number is sustainable.

In some fisheries, *observer coverage* is required. This means that a person independent from the fishing crew is on board monitoring and recording every creature that's caught. Some fisheries are moving to electronic monitoring, which provides high-definition imagery of everything that's caught for later analysis. Others require the fishing crew themselves to monitor and record what they catch. However, catching too many fish, undersized fish, or endangered species can lead to the fishery being suspended, which is not an outcome the fishermen want, so they may be tempted to simply not report any problematic catches. Still other fisheries require dockside inspections and monitoring, tracking not what's caught, but what's unloaded in port. These figures are reported to local and then national fisheries management authorities, who share annual catch statistics in their own reports. They also include these statistics as part of the biennial SOFIA (State of World Fisheries and Aquaculture) report released by the UN Food and Agriculture Organization.

In many fisheries, especially in the developing world, such monitoring is rare or absent because local governments just don't have the resources. One important research method that can help in these situations is the use of *market surveys*, which deploy trained researchers to regularly visit seafood markets where fishers land their catch to monitor and record what's caught. This is time-consuming, smelly work (follow scientist John Nevil on Twitter to see regular photos of fish he observes in the fish markets of the Seychelles) but it is vitally important.

>>MEET A SCIENTIST<<

The Role of Artisanal Fishing

Dr. Camila Cáceres, Researcher, Florida International University

The word *artisanal*, which is often applied to things like soap, jewelry, or beer, means *handmade*. *Artisanal fisheries* are ones that rely on a relatively low level of technology. They use small canoes or boats instead of larger, more powerful vessels. Many of these vessels are equipped with very simple, low-power engines, or no engines at all. Artisanal fisheries also employ traditional fishing gear such as spears or hand reels. These fisheries represent an important socioeconomic and cultural aspect of coastal communities, particularly in Latin America, Asia, and Africa. Unlike large-scale fisheries, artisanal fisheries do not have specific docks or landing sites to unload, register, and sell catch. Because artisanal fishers usually leave and return to their homes while fishing, there is little to no data about artisanal landings. Also unlike large-scale com-

Dr. Camila Cáceres. *Courtesy of the subject*

mercial fisheries, artisanal fisheries are of important social and cultural value to their communities. Both the fishing techniques they employ and the catches they take in have ancestral and historical significance to the fishers beyond either the physical or commercial sustenance they offer.

I conduct in-person interview surveys with fishers to study shark catches in artisanal fisheries. My research is focused in the Caribbean, Colombia, Trinidad and Tobago, Guadeloupe, Martinique, and the Florida Keys. Given the cultural differences, language barriers, and distance between these sites, I decided that interviewing and connecting with fishers would be better face-to-face instead of over the phone or with an online survey. When doing this work, I typically spent six to eight weeks in each country, living in a fishing village and driving around getting to know the islands as I looked for fishers to interview. Since artisanal fishers often leave to fish at dawn and are too busy with preparations to be interrupted, I waited for them to come back in the early afternoon to interview them while they cleaned their catch. I was incredibly surprised at how welcoming all the fishing communities were and how keen they all were to talk to me and share their thoughts. Now I realize how passionate and well-informed fishers are and how eager they are to have their voice included in science and conservation.

I have found that fishing communities are incredibly close and supportive of each other. If a fisher has a bad day, he or she can count on neighbors to share some of their catch to feed his or her family. Regardless of how fishing conditions change day to day, there is a safety net within these communities so that everyone is taken care of. On a day a fisher caught a bunch of tuna, the whole town celebrated his success with him. The commitment to family and community above all was inspiring and heartwarming. In my experience, it is difficult to find such support in "highly developed" urban centers. Supporting sustainable and ecofriendly practices in fishing is vital because many communities depend on seafood as a main source of both income and protein. Not eating seafood is not a realistic solution for most of the world; therefore, we have to be conscious to support well-managed fisheries.

Such surveys have led to crucial discoveries, including the detection of a shark that hadn't been seen in over a century. Scientists were shocked to discover specimens of the smoothtooth blacktip shark, a Critically Endangered species not seen in over a century for sale in a fish market. They never would have found this shark doing small-scale scientific sampling, but fishermen, out in force, caught the rare animals. I've never done a market survey for research, but love to visit local fish markets whenever I travel. (Special shout-out to my mom, the only member of my family willing to enter a very smelly fish market in Split, Croatia, on a recent family trip.) (See "Meet a Scientist: The Role of Artisanal Fishing" in this chapter.)

Such work also takes place at shark fin markets. DNA forensic tools can be useful to confirm which species of shark a certain fin came from (see "Meet a Scientist: Shark Fin Forensics" for more information).

>>MEET A SCIENTIST<<

Shark Fin Forensics

Dr. Diego Cardeñosa, Postdoctoral Associate,
Florida International University

Population genetics and DNA forensics are highly valuable tools that can help solve shark conservation issues. By knowing where wild populations occur and how much differentiation there is between them, we can propose management actions, assess the health of populations through genetic diversity, and even have an idea of how many individuals exist in a given population. Moreover, we can use that

Dr. Diego Cardeñosa (above). *Courtesy of the subject*

information to trace back shark fins from large trade centers to a population/region of origin to prioritize conservation efforts in those specific regions that supply markets with large volumes of fins. We are starting to track the fin trade in this way, and the identification of supply-chain starting points is a key piece of information that we did not have in the past.

To isolate DNA from individual cells, all you need is a piece of tissue from the shark or ray you want to study. Once you've extracted the cells, you can target informative genes and sequence them, then compare DNA sequences between individuals from different species and populations. For example, by looking at differences between the DNA sequences of separate species, we can design tools to detect illegal trade in sharks and rays and enhance the detection capabilities of border control personnel. This field of research, which will keep providing valuable information to protect these highly threatened species, has a very promising future.

Some of the most important and significant work we have done is to assess for the first time the full species composition of the largest shark fin markets in the world. We have analyzed over 10,000 samples to create a clear picture of which species are traded and in what proportions. This information has given us the opportunity to assess compliance with international trade regulations around the world. Like I said before, it's also allowed us to identify major supply chain starting points for some of the most threatened shark species and build capacity in these regions to improve the enforcement of conservation laws. These shark fin market surveys are optimal thermometers to detect changes in trade and fishing dynamics around the world. They also let us assess the efficacy of certain management actions at all governance levels and evaluate whether changes and improvement need to be put in place. This has been very exciting work that has already increased international protections for many species.

Human Dimensions

> Even though human pressures are one of the most important factors in the collapse of top oceanic predators, the social science of shark fishing has not kept pace with the biophysical science. Such a gap highlights the need for marine social science.
>
> —Key conclusions from "The Social Oceanography of Top Oceanic Predators and the Decline of Sharks: A Call for a New Field," a 2010 paper by Peter Jacques that created the field of human dimensions of shark conservation

There is increasing recognition among conservation biologists and natural resource managers that social science, the study of humans, is a vital piece of the conservation puzzle. None of those laws and regulations we talked about in chapter 6 cause sharks to change what they do; they only influence how humans can interact with sharks. It's therefore important to understand what humans are doing with sharks and why. By gaining insights into how humans think about sharks, we can begin to devise possible solutions and alternative approaches. This is known as the study of stakeholder knowledge, attitudes, and practices, or the *human dimensions* of wildlife management. Examining how humans interact with other species requires the incorporation of social science or interdisciplinary methods, including those I've been using in my research for years (see "Meet a Scientist: The Human Side of Shark Conservation Science" in this chapter for more information).

Some of the most common methods incorporated from the social sciences are surveys, interviews, and focus groups. These entail identifying a group of people who know something that you want to understand, asking them questions about that subject, and recording and analyzing their responses. I've done surveys and interviews of shark researchers, environmental advocates, and fishers; each of these groups is an important part of the big picture of shark conservation.

I've also tracked how information that could help shape stakeholders' knowledge of and attitudes toward different conservation policies is shared, as well as the rise of misinformation and disinformation shared through those information pathways.* I've used methods like *media*

---

*Misinformation is information that is simply incorrect, but not malicious. Disinformation is wrong information shared intentionally to cause confusion or disruption.

content and discourse analysis (obtaining every news article written about a subject and systematically evaluating what information is shared) and *virtual ethnography* (analyzing the postings on, say, a virtual message board used by the shark fishing community in Florida to get a sense of the community's knowledge, attitudes, and practices).

Though the human dimensions of shark conservation are a long-iden-tified research priority, this kind of research hasn't been sufficiently em-phasized. A 2020 analysis I performed found that out of nearly 3,000 presentations given at the American Elasmobranch Society conference, just 21 involved social science methodologies, and 8 of those were mine.

>>MEET A SCIENTIST<<

The Human Side of Shark Conservation Science

Hollie Booth, PhD Researcher, University of Oxford;
Sharks and Rays Advisor, Wildlife Conservation Society SE Asia

When I tell people that I work in shark conservation most of them imagine that I spend all of my time in the ocean swimming with sharks. The reality is that I spend most of my time talking to people, looking at dead sharks in fish markets, or sitting at my laptop analyzing data on fisheries and trade, then writing up policy recommendations. I also spend a lot of time working with early career researchers (ECRs) to provide mentoring and training and to work on joint research projects.

The greatest hazards to nature and biodiversity are driven by human choices and actions. In the case of sharks, the biggest threat to their survival is fishing, in some cases in targeted fisheries (mainly for their fins), but most frequently in fisheries for other commonly consumed seafood, where sharks are caught as bycatch. Therefore, effective shark conservation ultimately requires understanding and changing the human behaviors that drive shark overfishing, from at-sea fishing practices to consumer decisions. Just studying sharks

can give us lots of useful information about their biology and ecology, but social sciences can help us to design policies and interventions that ultimately reduce their capture and mortality in fisheries. In addition, since fisheries are important for people, both for income and food, we need to make sure that conservation policies and activities are feasible and ethical. That way, people can continue to benefit from the ocean in a sustainable manner, which can maximize positive outcomes for everyone.

My favorite parts of my job include chatting with fishers and learning about their lives and stories. If we're going to change fisher behavior, we first need to understand it; to convince people to open up to us, we must approach them with respect and an open mind. Spending time in the field is always fun and fascinating. People are kind and friendly, and I get to expand my perspectives, learn about new cultures, and hear some fascinating stories!

Hollie Booth (above). *Courtesy of the subject*

Emerging Research Priorities: What's Next?

At the start of the COVID-19 pandemic and associated lockdown, I realized that I probably wasn't going to get to do a whole lot of field research anytime soon. With the support and assistance of some colleagues, I pivoted back to something I'd wanted to do for a while: a nationwide survey of all relevant experts to determine what the next generation of research priorities should be for scientists wanting to study threatened species of sharks and rays in US waters. This research project had several advantages, not least of which being that I could do it safely in the middle of a pandemic because the entire process can be managed online.

Working with a graduate student and a team of colleagues, we identified hundreds of United States–based experts who have studied threatened species of sharks, work in fisheries management, work for the fishing industry and have participated in management discussions, or who work in environmental advocacy related to these issues. We sent these experts a survey asking which species they thought would require more research and management attention in the next decade. We also asked them to identify the most important unanswered research questions related to shark and ray species in US waters. The responses about research priorities cover a broad swath of conservation science methods and species. This paper (which as of this writing has not been published) will be a great resource for anyone curious about what's new and what's next in threatened shark conservation management in the United States.

On Becoming a Scientist

The sea was angry that day, my friends, like an old man trying to send back soup in a deli.

—George Costanza (Jason Alexander) pretending to be a marine biologist on *Seinfeld*

Some of the most common questions I'm asked by my social media followers are about career advice, and to be honest, I don't like answering these questions. There are several important reasons for this. First and foremost, as of this writing, I don't really have a traditional scientific career. I'm not a tenure-track professor at a university, so I certainly don't have any particularly useful advice on how to achieve such a position. Even if your goal is to become a staff scientist, not someone running their own lab, my path is far from the only path, and it may not be the best one. What worked for me won't necessarily work for others, and what didn't work for me might work wonderfully for someone else. Additionally, it's not a good practice to give potentially life-altering advice to people I don't know; without having any idea of a person's strengths, deal-breakers, or how far they can push outside their comfort zones before they have a panic attack, I'm unlikely to be useful.

That said, I'm happy to offer basic advice about how to work as a shark researcher without getting into specifics. Almost every professional marine biologist works for a university or for a government natural resources management agency like NOAA. A smaller but increasing number work for environmental nonprofit groups, where they either perform conservation research related to that NGO's goals and priorities or serve as technical advisors who assist the NGO's advocacy staff. People who take others scuba diving or fishing for a living but do not collect any kind of scientific data with the goal of answering a research question or contributing to a scientific publication are not working as marine biologists, though they may have some education in this field that helps them to find cool stuff to show their customers.

You'll almost certainly need some graduate school to work as a shark scientist. Depending on what kind of job you want, you may need to earn a master's degree or a PhD. A doctorate takes much longer to get and opens some doors, but some jobs don't require one. If you want those jobs, you're wasting a lot of time and associated earning potential by getting a PhD. But if you want to be a professor, run your own lab in academia or the government, or get a senior-level research role at a non-profit, you'll almost certainly need a PhD. As an undergraduate, you'll need to major in something science-y, though it need not be marine biology specifically (I majored in biology).

I also can't help you figure out the best school to attend for marine biology because there isn't one. If you aren't yet in college and are considering that step, I'd advise you to choose an undergraduate institution that offers research experience beyond cleaning lab glassware. It's good to have a sense of what kind of research you want to do going in because if you continue on with your studies, the graduate program you'll apply to will be different depending on whether you want to study age and growth or human dimensions of wildlife management. For instance, when contacting potential supervisors, it's much more useful to say something specific like "I am interested in studying the impact of PCBs [a man-made chemical that can be harmful when it leeches into the environment] on the reproductive system of blacknose sharks, which I know your lab is working on" than to say something vague like "I love sharks and want to save them." However, saying "I want to join your lab to study the impact of PCBs on the reproductive systems of blacknose sharks" is not a particularly helpful thing to say to someone whose lab does nothing of the sort, so make sure to tailor your job inquiries. By the way, some of the "famous" shark researchers may receive 100 or more email inquiries from prospective graduate students in a year when they only have 1 or 0 spots for new graduate students, so you should cast a wide net and apply to multiple institutions.

A great place to meet lots of potential supervisors at once is a scientific conference where many researchers in a given field gather together. For my readers in the United States, check out the American Elasmobranch Society (AES) meeting.

I'd be remiss if I didn't stress that while I love many of my scientific colleagues, our field has major problems with racial and gender diversity. The AES, for example, had 444 members in 2017, of which just three identified as Black or African American. The field also has a culture of pervasive sexual harassment and casual misogyny, which we are belatedly beginning to address. There are links in the further reading section on the book's website that address these and other issues.

Fortunately, these problems have solutions. In the summer of 2020, an important new group called Minorities in Shark Sciences (MISS), was founded. MISS focuses on improving representation, diversity, and inclusion in my field (see "Meet a Scientist: Working to Make Our Field More Inclusive for All" in this chapter for more information).

>>MEET A SCIENTIST<<

Working to Make Our Field More Inclusive for All

Jasmin Graham, Project Coordinator, MarSci-Lace;
President and CEO, Minorities in Shark Sciences Inc.

MISS (Minorities in Shark Sciences) was formed after Amani Webber-Schultz, Carlee Jackson, Jaida Elcock, and I found each other on Twitter during Black Birders Week through the hashtag #BlackInNature. We jokingly tweeted that we should start a club; after reading our posts, Dr. Catherine Macdonald of the Field School reached out to us and said that if we wanted to form some kind of group, we could host an event on the school's research vessel. We started thinking about it and realized that this was something that could really impact how the science world views women of color in marine science and how women of color view themselves, so we decided to launch MISS and try to fundraise to cover the travel expenses of participants.

Our goal is to be seen and to take up space in a discipline which has been largely inaccessible for women like us. We strive to be positive role models for the next generation. We seek to promote diversity and inclusion in shark science and encourage women of color to push through barriers and contribute knowledge in marine science. Finally, we hope to topple the system that has historically excluded women like us and create an equitable path to practicing shark science. We believe diversity in scientists creates diversity in thought, which leads to innovation.

Our plan is to use our collective voice to speak out against biases and discriminatory practices prevalent in marine and shark science. We also want to hold organizations accountable for gatekeeping and creating barriers for people of color trying to break into the field. Finally, we plan to do whatever we can to provide mentorship, support, networking opportunities, and professional development to our members.

Jasmin Graham with a sawfish caught as part of her master's research, which was conducted under ESA Permit 17787. *Courtesy of the subject*

If you want to experience life as a shark scientist in a safe and inclusive environment, my friends at Field School (see "Meet a Scientist: The Limits of Wildlife Tourism and a Safer Path for Shark Science Education" in this chapter for more information) have a shark field research skills course a few times a year. Attendees live on a research vessel, learn from shark researchers (including [sometimes] myself), and get to practice the skills they learn on some sharks.

>>MEET A SCIENTIST<<

The Limits of Wildlife Tourism and a Safer Path
for Shark Science Education

Dr. Catherine Macdonald, Founding Director, Field School;
Lecturer and Master of Professional Science, Marine Conservation;
Track Coordinator, Rosenstiel School of Marine and Atmospheric
Science, University of Miami

On Wildlife Tourism As I always tell my students, any time something seems black and white to you (all good /all bad), you probably just don't understand it well enough yet. You should be skeptical whenever someone offers you a silver bullet solution that promises to fix environmental problems without costing anything. It's more likely you just haven't noticed the costs, generally because they fall on someone else.

How does this apply to wildlife tourism? At its core, tourism isn't a practical solution to shark conservation because:

1. Most sharks aren't considered desirable for tourism. Tourists aren't lining up to swim with a two-foot-long cat shark, or to dive in frigid waters, or to travel long distances offshore. Sharks that can support a significant tourism industry are generally large-bodied species that live in shallow coastal waters with good visibility. That's a pretty small proportion of all sharks.

Dr. Catherine Macdonald (above). *Courtesy of the subject*

2. Tourism doesn't necessarily lead to conservation. There's a lot of news coverage talking about sharks being worth more alive than dead because of tourism, but thinking about total economic value isn't adequate; you also need to consider how that money is distributed. Fishers, fish processors, and fish vendors aren't necessarily all going to be readily able to transition to a career in tourism, and they may not want to. At the end of the day, people can't eat tourism, and a ton of tourist operations don't do a lot to benefit local communities where they're located, either. The economic incentives created by tourism aren't necessarily simple or straightforward.

On Field School's Mission Field School is a marine science education organization offering classes for students who want to learn marine science field skills, especially those related to shark research. I co-founded the school to address a real lack of opportunities to learn these kinds of skills in safe, supportive, structured environments. In many cases, marine field skills aren't formally taught, and students are left to figure things out for themselves. As a teacher, I realized that wasn't great for students or sharks. I wanted to create a better way to learn than what was available in the system I came up in, one that would be friendlier to people from historically excluded groups in STEM.

I hope that this brief introduction to how scientists gather data to help protect threatened sharks, and how people can become scientists and contribute to these studies, gives you a broader appreciation of the complexity and nuance of the work we do.

9 >> How Are Environmentalists Helping Sharks?

By 2025, the conservation status of the world's shark and rays has improved, declines have been halted, extinctions have been prevented, and commitments to their conservation have increased.

—Goal statement of the Global Sharks and Rays Initiative 2015–2025

Scientists and scientific data are vital to help plan effective conservation policies, but merely publishing data in peer-reviewed scientific journals does not, in and of itself, result in policy change. We need partners who can help craft policy initiatives rooted in the results of our research. This is where environmental advocacy becomes crucial to help implement applied science in practice. So let's talk about the work environmentalists do, and in the course of that discussion, it will be my pleasure to introduce you to the shark conservation movement! While I've worked closely with some environmental advocacy organizations in my capacity as a scientist, I'm not as steeped in the inner workings of this world, so this chapter is by necessity a little more general than the previous chapter on scientific research.*

When considering the missions and actions of conservation organiza-

---

*As a reminder, this book focuses on science-based shark conservation, including the scientific evidence for why we need to protect sharks and which sharks to prioritize for protection. I also support the data-driven, evidence-based policies that have been shown to be most effective at protecting sharks. There are arguments for protecting sharks that don't have their roots in science, and these aren't wrong, they're just not what I'm interested in and not what this book is about.

tions, it's important to distinguish between these four key terms used to describe their efforts: *public education, advocacy, activism,* and *lobbying.* All of these concepts have a role to play in changing the world and solving problems, and the boundaries between some of them can get blurry, but they are not interchangeable synonyms.

Let's start with a relatively straightforward concept that's familiar to most people, *public education.* This involves either experts on a topic—or non-expert supporters or opponents of a policy outcome related to that topic—speaking to members of the public. There's usually some component of *awareness raising* associated with public education; the goal is to make people aware that a problem exists, as well as to educate them about various possible solutions.* Many environmental nonprofits include a public education component; some focus exclusively on this approach.†

While we're talking about raising awareness, I have to briefly interject that "raising awareness" of incorrect information in support of policy solutions that aren't backed by evidence is, though quite common, not helpful. Without a firm grasp on accurate information about shark conservation, starting your own public education campaign is perhaps not the best idea, In fact, one of the reasons I wrote this book is to try to help prevent this kind of mistake.

The next two terms, *advocacy* and *activism,* are often used interchangeably, but that isn't really correct. So, you might wonder, what *is* the difference between advocacy and activism? The waters here are so muddy that the sources I used to look up definitions for writing this section don't agree with each other. Everyone agrees that there is a difference, though, and that the difference matters. In the science-based shark conservation space, the distinction is often a question of methods. You're more likely to see an *advocate* speaking at a prearranged time for public comment in a natural resources management agency hearing.

---

*Once a public education campaign focuses exclusively on one solution, rather than sharing the message that in reality multiple solutions exist to most environmental problems, the borders between public education and activism get blurry fast.

†By the way, if you want some great free shark science materials that are suitable for a variety of grade level students, check out the organization Sharks4Kids.

An *activist*, on the other hand, is more likely to stage a public protest outside of the building. A mainstream environmental nonprofit like the Ocean Conservancy employs people in roles as advocates, whereas a *direct action* group like the Sea Shepherd Conservation Society, the star of the Animal Planet show *Whale Wars*, is made up of activists who have been known to do things like ram whaling vessels.

While activism need not be extreme, violent, or controversial, advocacy is essentially *never* any of those things. Organizing a sedate letter-writing campaign can absolutely be a form of activism, for instance. But you may find that people who consider themselves advocates feel insulted if you call them activists because that word is so loaded. Activism has come to be associated by many with more extreme measures which (at least in theory) don't require participants to have as much formal education and experience as knowledgeable advocates. Similarly, you may find an activist offended to be called an advocate because they believe that dramatic, flashy activism is more effective. They may feel that the crisis we face is an emergency and that this is no time for waiting your turn to speak at a hearing. Scientists may sometimes operate in the sphere of advocacy, but activism, as I indicated previously, can be considered inappropriate and unprofessional for scientists to engage in. On the other hand, most scientists would have no problem participating in a public education effort that consisted of, say, speaking to a middle school science class about what life is like as a marine biologist, while noting that sharks are in trouble and describing some existing policy solutions.

All of these terms need to be distinguished from *lobbying*. Though the word itself has many definitions, in environmental advocacy circles lobbying usually refers to a very specific subset of advocacy targeting elected legislators. Asking NOAA to reduce a fishing quota is not lobbying. A concerned member of the public writing to their congressman asking them to support a bill is not lobbying. However, an employee or an environmental group asking that same congressman to support that same bill can be a form of lobbying. Caution is advised when using this term because in the United States people are required to register as lobbyists, and environmental advocates are often *not* registered lobbyists (though their nonprofits may hire registered lobbyists to speak on their behalf). If you say that someone who works for an environmental group is lobbying and they are not, you may receive unexpectedly rapid and forceful pushback. Even if you meant no harm, you've declared that they are doing something illegal. Sometimes the difference between activism and lobbying comes down to who you're asking to make a change or who employs you while you're making that ask.

As for me? I've done a lot of public education; heck, this whole book is a public education project. I've never engaged in lobbying, and I've rarely participated in activism. Occasionally I've engaged in advocacy—the paper I contributed to the journal *Marine Policy* stating my objections to the proposed shark fin trade bans are advocacy, for example, though my coauthor, Bob Hueter, and I were careful to frame these objections in terms of the available data and evidence. I've also worked behind the scenes to assist environmental groups engaging in advocacy either by sharing their content with my social media audience (which also supports public education) or helping to craft their messages and ensure that these messages are based on the best available scientific facts.

The groups that I'll mention in this chapter mostly do what I'd call advocacy, with a little activism mixed in.

Some Great Shark Conservation Nonprofits That You Should Know About

Our mission is to provide leadership in advancing sound, science-based local, national, and international conservation policies through collaboration with a diverse array of organizations and decision-makers.

—From the mission statement of Shark Advocates International, my go-to shark conservation nonprofit

As the conservation of sharks becomes both increasingly recognized as a science-based priority and increasingly popular among concerned members of the public and donors, there has been an explosion in the number of environmental nonprofits that focus on shark conservation advocacy issues. As part of my postdoctoral research, I identified 78 environmental nonprofits across the English-speaking world that either exclusively focus on sharks or have at least one active shark conservation campaign. These organizations range in size from local-scale operations with just one part-time employee to massive international nonprofits with over 1,000 full time employees based in dozens of countries. Some are "all sharks all the time," while others engage in a variety of ocean conservation campaigns including but not limited to shark conservation. Still others have land and sea conservation missions. These environmental organizations also range widely in both scope and methods, with some focused on enacting the evidence-based, data-driven conservation priorities I've mentioned throughout this book. Some, however, appear to operate with distinctly different standards, while others very obviously do not focus on evidence-based data-driven conservation priorities.

Troublingly, my research identified a small but vocal group of nonprofits that do not engage with scientists, scientific facts, or anything that could reasonably be described as evidence. Members of these nonprofits regularly say things that are obviously factually incorrect in support of policy preferences that are not shared by scientific experts. (This problem is distinct from the one created by animal welfare groups who don't base their arguments on science, but also don't claim that they do. The questionable groups I identified instead claim to use science, but then say things that are not scientifically true.) Although my social media followers may regularly see me griping about nonprofits that misrepresent themselves (and shark science!) this way, I do want to stress here that there are some wonderful advocacy groups that I enthusiastically support. And I am far from the only scientist who sees great value in science-based advocacy. In my 2016 survey of American Elasmobranch Society members, many AES shark scientists reported that, despite concerns about accuracy or focus in some corners of the conservation world, they support at least one environmental nonprofit financially or by donating their expertise when needed. The conflicts between the mainstream and extremist wings of the conservation movement are real and significant, but there's also a lot of effective collaboration that doesn't generate sketchy headlines or outrage clicks.

The purpose of this chapter is not to introduce you to every single organization out there, but to highlight some great examples of non-profits whose work I support. I also aim to give you a sense of the range of groups that exist. It's not the goal of this book to call out all the bad actors in the shark conservation space by name, but I'll try to highlight some behaviors that raise red flags. I'll also offer some tips for spotting a reliable organization you can trust. I do want to warn you that this chapter may contain out-of-date information if you're reading it years after it was published because environmental nonprofits can change their areas of focus. For example, if you had asked me two years ago if I thought the Pew Environment Group would phase out their extremely high-profile sharks team, I would have had a good chuckle—yet that's exactly what they've done as part of a renewed focus on international fisheries management and establishing new marine protected areas.

This book's website mentions ways you can learn more by following some of these nonprofits and their expert employees on social media and by subscribing to their newsletters. It also includes links to donate to support their valuable work. (During edits of this chapter, my editor asked me if promoting a donation link was a type of advocacy. According to some definitions of advocacy, the answer would be yes; according to others, it would be no. I told you that this stuff can be confusing!)

Shark Advocates International

Throughout my career, my main mission has been to show up and speak up for elasmobranchs. I serve on every government advisory panel that will have me.

—From a profile of Sonja Fordham, Shark Advocates International founder and president, in *Save Our Seas* magazine

I mentioned that Shark Advocates International, led by conservation hero Sonja Fordham (see "Meet a Conservationist: Showing Up to Speak for Sharks [and Skates and Rays]" in this chapter for more information), is my favorite shark conservation nonprofit. It's one of very few nonprofits whose work I can recommend with no qualifications attached.

SAI is a small, targeted, and nimble operation, which means that if you choose to support them financially, your donations go a long way. As indicated in the quote above, Sonja Fordham, the organization's founder and president, goes wherever there's a government discussion about a shark conservation issue, sometimes as part of a big coalition of many nonprofit groups and other times as literally the only person at the whole meeting speaking on behalf of shark (or skate, or sawfish) conservation. When there's a committee or panel that doesn't have a formal place for a member of the environmental community, she speaks in her capacity as a concerned member of the public, an option open to us all, but that few of us take advantage of. As the cliché goes, decisions are made by those who show up, and Sonja Fordham shows up.

You might find her speaking in Geneva in support of listing mako sharks at CITES, in Rome attending a UN planning meeting on global sustainable fisheries issues, or in a conference room at the Holiday Inn–Miami International Airport discussing sawfish conservation in Florida. She's also the deputy chair of the IUCN Red List's Shark Specialist Group and the chair of the American Elasmobranch Society's conservation committee. On top of all that, she has a standing policy that if a scientist, especially a graduate student, has a policy question related to their work, she's happy to answer it. She commits her time to this in order to make sure that we scientists know what we're talking about when we venture into the policy world.

What does SAI actually do? A lot of the organization's job is essentially sharing a pro-conservation message with decision-makers through already-existing (though often unused) channels for that information. This usually means talking with government fisheries managers rather than Congress, for example. SAI contributes to discussions at some of the natural resources management meetings we talked about in chapters 5 and 6. SAI also works with coalitions of other nonprofit groups and writes position statements outlining why the coalition believes policymakers should support or oppose a proposed management measure.

To give you a further sense of the scope of issues SAI works on, the list of position papers on their website includes:

- A 2012 letter urging fisheries management authorities in New Zealand to ban the catch of endangered oceanic whitetip sharks in their Pacific fleet;

- A 2012 letter strongly recommending that the Western and Central Pacific Fisheries Commission ban setting tuna nets around whale sharks (this is basically the same issue that led to the creation of the dolphin-safe tuna designation, but as it pertains to whale sharks);

- A 2013 public comment read aloud at the Atlantic States Marine Fisheries Commission meeting requesting them not to enact a 12% fin to carcass ratio for smooth dogfish;

- A 2014 letter calling on the government of Western Australia not to cull great white sharks in a misguided beach safety program;

- A 2016 position statement for the international representatives of the Northwest Atlantic Fisheries Organization urging them to enact a *retention ban* on long-lived Greenland sharks. Such a ban requires fishers who catch even one shark to let it go, even if it's dead. This policy is distinct from regulations that encourage fishers to let live sharks go, but permit them to land and sell dead ones. As you might imagine, such permissive policies encourage fishers to engage in creative machinations to report that a live shark was actually dead.

You could get a pretty solid lesson in the history of shark conservation over the years simply by examining SAI's written statements in support of or in opposition to these issues. SAI has been involved in just about every major science-based conservation policy discussion since my parents said I was too young to watch *Jaws*.

>>MEET A CONSERVATIONIST<<

Showing Up to Speak for Sharks (and Skates and Rays)

Sonja Fordham, Founder and President,
Shark Advocates International

Shark Advocates International, a project of the Ocean Foundation, was founded to advocate for science-based shark and ray conservation policies with fisheries and wildlife agencies in the United States and around the world. We focus on stopping overfishing and work wherever possible as part of coalitions. My 30 years of extensive experience in shark conservation policy and solid scientific networks allow me to provide leadership in several arenas, particularly fisheries management. We expend a good deal of effort encouraging scientists and concerned citizens to add their voices to balance fisheries debates. We put a premium on perseverance and accuracy and pride ourselves on speaking up, especially in tough arenas. Our small size and deep expertise allow us to react nimbly to shark conservation events.

At the moment, we are embroiled in a long fight to secure an overdue international ban on the retention of endangered North Atlantic shortfin mako sharks. This is a critical measure promoted by scientists to prevent population collapse and reverse serious decline. At the moment, the biggest obstacles to achieving this agreement are the positions of the European Union and the United States, despite both countries' solid records for championing science-based shark conservation measures in the past.

Sonja Fordham (above). *Courtesy of the subject*

Sharks stand out for being treated as both wildlife and commodities. Effective conservation has been plagued in almost every fishing nation by the chronic disconnect between fisheries and wildlife agencies. I think it's vital that conservationists help bridge that gap by engaging not only with environmental officials but also within the fisheries world. Fisheries arenas are generally more taxing and unpleasant for environmentalists now compared to wildlife protection meetings, but it doesn't have to be that way. I believe fishery managers would get more accustomed and open to conservation concerns if we—like the fishing industry—maximized our opportunities to directly explain our perspectives (politely, of course).

I continue to ask concerned citizens to let policymakers know that they care about shark conservation. Whether it's about a local problem or a global initiative, resource managers and elected officials need to hear from the public, not just fishing interests. Although we've made amazing strides, sharks still garner a lot of negative feelings and remain relatively low priority; sustained public support is vital to continuing progress. It's important to stress that people need not be experts on a shark conservation issue to weigh in; genuine concern is sufficient, meaningful, and potentially quite effective. It's best when such expressions reflect at least some of the person's own voice and perspective and when they sustain their engagement over time.

Conservationists rarely get all we argue for, much less what is needed to truly minimize threats, and yet real benefits can come from interim steps. We feel pressured both to be upfront about how bad things are and to demonstrate concrete success from our efforts. I try to celebrate the milestones that I judge as true advances, but never without pointing to the necessary follow-through actions and immediate next steps.

Lately, much of SAI's focus, in partnership with the Shark League for the Atlantic and Mediterranean (a coalition of other nonprofits), has been on Atlantic mako shark conservation. These sharks are sought after for not just their fins, but their meat. You used to be able to get mako shark steaks at the seafood counter at the Giant Eagle grocery store near my childhood home in Pittsburgh. Recent evidence has shown that these sharks are not only severely overfished but that their populations have declined so much that they qualify as IUCN Red List Endangered. The Shark League's goal is a total ban on landing mako sharks, commonly caught as bycatch in Atlantic swordfish and tuna fisheries. Makos obtained CITES Appendix II listing, but the total ban was voted down at ICCAT, a tuna Regional Fisheries Management Organization whose purview covers shark bycatch. My extensive writings on the saga of Atlantic mako shark conservation have quoted both Sonja Fordham and the SAI's Shark League partners.

The Shark Trust

Our work is deeply rooted in science. Science removes guesswork. It enables us to make informed decisions so we can direct our efforts where it's most needed and get results.

—From the mission statement of the Shark Trust

The United Kingdom–based Shark Trust is a science-based environmental nonprofit that focuses on saving endangered shark species (such as angel sharks, basking sharks, and porbeagle sharks) and improving fisheries management for overfished species (such as mako sharks and blue sharks, as well as some smaller coastal species whose names you've probably never heard). They regularly collaborate with other nonprofits, including those mentioned in this chapter, and are another member of the aforementioned Shark League. They also do truly impactful citizen science work through the Great Eggcase Hunt, which encourages beach users to find and record shark and skate eggcases that wash up on beaches. They also engage in lots of great public education efforts.

Shark Trust is perhaps best known for their No limits? No future! campaign, which focuses on creating sustainable fishing quotas to end the unlimited fishing that threatens shark populations around the world. In addition to mako and blue sharks, this campaign's focus includes some species caught in coastal waters like tope sharks, smooth-hounds, and catsharks. In recent years, Shark Trust has concentrated largely on mako sharks. The conservation "bad guys" in the mako story are, surprisingly, Spain and Portugal, whose fishermen catch over 25,000 makos each year. This rampant overfishing is enabled by the United States and the European Union, which have blocked international mako shark conservation efforts. Makos aren't their only concern, though; their website mentions that in 2019, the Trust was instrumental in convincing world leaders to implement an international Regional Fisheries Management Organization catch limit for blue sharks in the Atlantic ocean.

The Shark Trust's work on angel shark conservation as a partner in the Angel Shark Conservation Network is also noteworthy. Because these flat-bodied sharks burrow in the sand, they get caught in fishing trawls that are dragged across the ocean floor. This has put them in a particularly vulnerable situation, making efforts to save them vitally important.

MarAlliance

Local empowerment is critical to build long-term commitment to the conservation of species.

—From the mission statement of MarAlliance

The vast majority of shark conservation nonprofits are based in Europe or North America, leaving important gaps in much of the global South that local branch offices of international NGOs don't always do a great job of filling. The MarAlliance is one of the most effective conservation nonprofits working in Central and South America, though they also have a project in the South Pacific. They engage in plenty of their own research and public education efforts, but what I most deeply respect about the MarAlliance (and why I want to highlight it in this book) is their committed outreach to fishing communities. When Dr. Rachel Graham, MarAlliance's founder and executive director, traveled to Brazil to give a keynote address at the 2018 Sharks International Conference, she brought a Belizean fisher partner with her so he could meet with scientists, conservationists, and fishers from all over the world.

As we've discussed, a top-down approach to conservation in which the government just says "You can't do this anymore" is less effective than a conservation initiative with community support behind it. This is doubly true when the government of a developing nation is being pressured by environmental activists from a wealthy nation, which happens all the time—an approach that reeks of colonialism and privilege. That's what makes the MarAlliance's efforts to work with, teach, and listen to fishers all the more important. They've not only trained fishers in how to safely release sharks they don't want to catch and how to fish in ways that are less likely to catch sharks, but have also taught over 100 fishers to be field research assistants on a variety of projects. Their fisher partners serve as great ambassadors to the rest of the fishing community, demonstrating that sometimes the messenger matters as much as the message.

Project AWARE

If you're a scuba diver, you might have heard of Project AWARE, which partners with the Professional Association of Diving Instructors (PADI) to give the global scuba community an easier way to join forces for ocean conservation. They've created some initiatives specific to divers, like implementing a code of conduct to make sure divers don't damage the ocean. They offer a series of speciality PADI certification courses that teach divers about ongoing ocean conservation issues. They also arrange citizen science opportunities for divers and organize beach clean-ups.

Project AWARE has hired professional environmental advocates who join in a variety of multi-nonprofit coalitions. These advocates have played key roles in successfully securing CITES listings of a number of shark species, as well as achieving victory in a years-long campaign to get the European Union to close loopholes in their shark conservation regulations. Interestingly (and somewhat controversially in some quarters of the environmental nonprofit space), Project AWARE recently decided to have their members vote on what issues to tackle next rather than identifying priorities and expert capacity on their own. Happily, sharks won the vote and remain one of the Project AWARE priority areas, at least for now.

Defenders of Wildlife

This report was produced in response to a petition received from Defenders of Wildlife on September 21, 2015 to list the oceanic whitetip shark as endangered or threatened under the Endangered Species Act. The National Marine Fisheries Service announced that the petition has sufficient merit for consideration and that a status review was warranted.

—From a NOAA fisheries status review of the oceanic whitetip shark

A curious aspect of the United States' natural resource management and endangered species conservation system is that while the government can sometimes do the right thing proactively, it often has to be nudged, petitioned, or sued to do the right thing. While any citizen has the right to petition the government, doing so usefully and effectively often requires some specialized knowledge and training in science or law. (Please keep in mind that this doesn't mean the government is inherently bad; some government scientists and managers I know are thrilled when they get sued because it means they'll be forced by court order to do a thing they wanted to do anyway but weren't able to due to lack of resources or political pressure.) One stellar recent example of this phenomenon is the welcome listing of the oceanic whitetip shark as Threatened under the United States Endangered Species Act, a listing that requires NOAA to generate a protection and rebuilding plan. This decision was made because Defenders of Wildlife, a group that usually focuses on terrestrial conservation issues but gets involved in the legal side of shark issues from time to time, wrote and submitted a petition explaining why whitetip sharks needed help that only a formal ESA listing could ensure. A similar petition for Atlantic mako sharks has been submitted, but no decision has been made as of this writing.

To be clear, a "petition" like this is not the ineffective "Ban baby sharks in jars" thing your dive buddy shares on Facebook. Defenders' petition to list oceanic whitetip sharks, which you can read on this book's website, is 111 pages long, took a team of PhD scientists and attorneys several months to write, and cites about 100 peer-reviewed scientific references. It systematically goes through requirements for ESA listing and provides concrete evidence that whitetip sharks meet those requirements. Crafting an effective petition involves an enormous amount of work and effort. It also requires specialized knowledge and expertise of the type that Defenders of Wildlife has, but that ordinary concerned individuals who want to help probably lack.

Some environmental nonprofit groups generate absolutely enormous quantities of Endangered Species Act petitions, including many for species that don't seem to need protecting. This is, at best, a waste of time. The rules are clear about what it takes to get ESA protections and which species do and do not quality. Targeted ESA petitions like Defenders' whitetip shark petition, on the other hand, can make a true positive impact.

Defenders also joins a variety of coalitions of nonprofits focusing on shark issues, perhaps most notably those dealing with the Convention on Migratory Species.

How to Find an Effective Conservation NGO to Support

Charity Navigator was founded by Pat and Marion Dugan, who were passionate about charitable giving but witnessed report after report of scandals.

—From the history of Charity Navigator, an impartial guide to nonprofits and how they spend their donor funds

This chapter has given you a brief introduction to the broad range of approaches that some environmental nonprofits take to changing laws and behaviors to protect sharks. However, these certainly aren't the only nonprofits out there. How can you tell if a shark conservation organization is worthy of your support?

Charity Navigator is a commonly cited and well-respected tool for evaluating nonprofits. As mentioned in the quote opening this section, there are indeed issues with scandal, fraud, and abuse in an alarming number of nonprofits. But it's important to understand that Charity Navigator only tracks things like how nonprofits spend their money and if they spend it on mission objectives versus perks for the organization's board members. This means that if a group delivers on their stated objectives, even if those objectives were bad, Navigator wouldn't flag them. They're concerned more with fraud and waste of donor funds, not whether a charity focuses on evidence-based outcomes. Therefore, digging a bit deeper than the information Charity Navigator can provide may be desirable.

Some of the questions I ask myself when choosing whether to support a nonprofit include:

1. **Have I ever heard of them?** If the answer is no, that may just indicate that they're new, which isn't necessarily a problem. Everyone has to start somewhere, after all. However, it may also mean that they have no track record of successful outcomes; if they had achieved many of their goals, I'd probably have heard of them. Of course, just because I've heard of a group doesn't mean I've heard positive things about them. In short, this question can be a useful "smell test" (an informal way of assessing "Does this seem credible to me?") even if it's not the ultimate deciding factor.

2. **Does the nonprofit have a website, and is it easy to determine what they do from looking at it?** If the answer is no, that isn't necessarily a deal-breaker, but an effective organization in the twenty-first century is very likely to have a website. Not being able to find one is a bad sign. A website full of vague platitudes like "We save the ocean" without any details is also a bad sign. By this point in this book, you know that details matter, and that there are many types of policies out there. What issue does the group focus on? Overfishing? Bycatch? Endangered species management? Habitat protection? Creating protected areas? If you can't tell after making a good faith effort, that is a red flag. If you can determine what issue they champion, what policies do they support to address it?

"Help us fight overfishing" could mean anything from promoting science-based fishing quotas (which I support) to ramming fishing boats (which I strongly oppose).

3. **Does this group work with other nonprofits?** Collaboration, especially with groups that have a great reputation for using evidence and science in support of successful outcomes, is a good sign. A total absence of collaboration could indicate a problem because it either means that no one wants to work with them or that no one else cares about the issue they're working on. It's unlikely that any issue that's really worthy of attention wouldn't attract any partners. All of this information should be very easy to find. If it isn't, that's another red flag for me. "We're proud to work with a big team to solve a thing we all agree is a problem" is certainly more credible than "I am the only person in the world who thinks this is a problem, and I'm the only person you can trust to solve it."

4. **How does the group engage with science and evidence?** Science isn't the only factor in conservation, and it's perfectly acceptable to base your entire argument on personal or cultural values. However, I personally am more moved by cases where evidence shows that a particular issue is critically important and demonstrates that a group's preferred method of handling the problem is an effective way to solve it. Some nonprofits, however, seem to want to ban certain practices simply because the group thinks they are bad, evidence be damned. A corollary issue is that if a nonprofit claims to use science to shape its policy preferences and does not employ scientists, work with scientists, or incorporate any actual scientific evidence into its advocacy or activism, that's at best a little disingenuous.

5. **What specific aspect of an organization's work am I being asked to support?** Some nonprofits, especially the larger ones, are actively engaged in many different campaigns at any particular moment in time. A group may be working on something I don't support while simultaneously working on something I think is ex-

tremely important. I'm personally quite willing to support some things that a group does, even if I don't support everything they do, but I usually try to offer qualified, specific support that can't be misconstrued.

I hope that this chapter has given you a brief window into the world of conservation advocacy and how it overlaps with—but is separate from—the worlds of conservation-relevant science and policymaking. If you are interested in getting involved with shark conservation advocacy, I encourage you to check out some of the resources and links on the book's website that can put you on the path. And on that note, there is still one incredibly important source of support for sharks that we haven't yet discussed: you!

10 >> How Can You Help Sharks?
(Dos and Please Just Don'ts)

If you find that your uninformed personal opinion disagrees with expert consensus on a technical topic, please consider that "All the experts know something I don't" is a more likely explanation than "I alone solved a complex problem that experts have long been studying."*

—A tweet by the author

N ow that you know why sharks aren't a threat to you, why sharks matter, and that many sharks face conservation challenges, I've done my job as an author. I hope the very next thing you'll wonder is "What can *I* do to help sharks?"

If you are a scientific or policy expert, I assume that you already know what you can do to help sharks (thanks for buying my book anyway!). But now I want to speak directly to folks who are aren't scientists, policymakers, or professional environmental advocates. I know that many readers of this book haven't devoted their professional lives to ocean science or conservation, but still truly care and want to do something to help. The good news is that there are many things that enthusiastic supporters of shark conservation can do to support sharks and the scientists, conservation advocates, and policymakers who are dedicated to helping them.

The bad news is that there are lots of people who want to help sharks, but without understanding the principles and practices of ev-

---

*This is my second-most-shared tweet of all time. (As of this writing, my most-shared tweet of all time points out that you can sing *COVID-19* to the tune of "Come On Eileen.")

idence-based conservation, they try to reinvent the wheel rather than listen to experts. Not only do these well-meaning folks fail to help sharks, but they can make it harder for my expert colleagues and me to do so. Their misguided efforts can take away time and attention from proven solutions that really could use your assistance. Even worse, there are some downright unscrupulous people out there who will try to take advantage of your desire to help through what can only be described as fraud.

Some people think my objections to non-expert solutions are just sour grapes or elitism. They accuse me of trying to discourage people from helping. Honestly, these accusations just don't make sense. Speaking on behalf of anyone who ever cared about a cause, I'd be thrilled if we saved the ocean. I wouldn't have to be a marine conservation biologist any longer; instead, I could just study pure shark behavior without worrying about how results affected extinction risk. Obviously, I've dedicated my professional life to preserving healthy shark populations, so of course I'm all in on any solution that can truly help. While it's true that scientists don't know everything, a data-driven, expert-backed solution is far more likely to be effective than one made up by someone who may feel passionately about the problem, but hasn't taken the time to learn important background facts. So what is—and what is not—going to help save sharks?

Submitting Formal Public Comments (and NOT Signing Silly Online Petitions)

If you live in the United States, you have an incredible opportunity to shape the course of environmental policy. Our natural resource management and endangered species conservation systems are participatory by design; any proposed changes to rules and regulations are publicized beforehand to provide enough time for concerned citizens to speak out for or against those intended changes. In fact, NOAA is legally obligated to consider and respond to formal public comments submitted through their official portal. Of course, this doesn't mean that they'll do whatever you say; they also have to consider and respond to people who disagree with you, and there are overall limits on what's possible. But this is a real opportunity. If you submit your comments through NOAA's portal, they are legally bound to consider and respond to your opinion.

To improve your chances of effecting change, I'd discourage you from writing in with vague feel-good platitudes like "Save the sharks!" As you've seen in this book, there is a wide range of things that can be done to save sharks. If you aren't specific in your ask, our friends at NOAA get a little more wiggle room to ignore your request. Instead, you'll have to write more detailed things like "I oppose amendment 16B because it doesn't appropriately consider bycatch. Instead, I support amendment 14A, which will have better outcomes for this threatened shark species." It's unlikely that a passionate member of the public who doesn't have relevant knowledge, expertise, education, or training would understand that level of nuance. This is why it's important to listen to experts who know what's really going on. Reading this book is a good way to start developing more informed conservation opinions and strategies. I also share ways to help on my social media accounts whenever they come up, which is usually a few times a year. Other organizations put out specific calls for help as well; as I've mentioned in previous chapters, you can follow them on social media or sign up for their email lists.

Conversely, not bothering to comment on proposed policy can be damaging. True, public comments aren't the be-all and end-all of ocean conservation policy in the United States. It's unlikely that a swarm of public comments would generate enough momentum to allow a wildly unrealistic proposal to pass. However, there have been many proposals over the years that have come really close to getting the desired outcome, and a few more public comments favoring that outcome may well have tipped the scales. Over the last five years, there have been about a dozen significant shark management policy decisions open to public comments. It's not uncommon for these discussions to pass by with no more than a small handful of clear and specific public comments in support of the pro-conservation option.

Let's contrast the potential impact of these public comments with the online petitions that my social media followers regularly see me gripe about. Perhaps the most important difference between online petitions and public comments is that even if such a petition is well-written enough to avoid the other issues I'm about to raise, NOAA is not legally bound to look at these potentials, no matter how many signatures they get. Several NOAA natural resource manager colleagues have told me that they usually don't look at these amateur petitions at all—and even when they do see them, as few as *five public comments* may factor more strongly into a NOAA decision than *five thousand signatures* on an online petition. It takes a lot more effort to craft a public comment than to sign an online petition, and NOAA decision-makers know that just as much as you do (now).

The most common platform for these petitions is Change.Org, which is integrated through many leading social media channels to allow easy spread through your networks. This means that, for example, you can sign a petition directly from Facebook after seeing your friends sign it, and you can easily share the link with your other friends. The websites behind these petitions advertise that they make it easy for anyone to propose a solution and change the world. This sounds great, but the problem is not everyone knows how to effectively propose solutions. Crafting a catchy conservation slogan that goes viral on social media is not the same thing as proposing a detailed, data-driven, evidence-based conservation policy. If we care about successful outcomes, fewer people should make online petitions, not more. As I wrote in a 2013 blog post entitled "How to Make a Completely Useless Online Petition in 5 Easy Steps," "A well-written petition can be an important tool for helping to shape policy, particularly when used as a part of a larger well-organized advocacy strategy. Many online petitions, however, are so badly written as to be ineffective or even counterproductive."

This doesn't mean that petitions don't have their place. Petitions can indeed be useful if they're part of a thorough, well-researched conservation campaign. A science-based nonprofit may use them to be able to assert something like "Our detailed analysis of the data shows that this is the right way to protect endangered species—and by the way, 20,000 Americans agree with us and signed our petition." But petitions are extremely unlikely to be useful on their own, especially if the folks making them have no knowledge, education, training, or experience related to conservation advocacy or the science behind policy solutions.

Some particularly egregious examples of useless online petitions I've seen over the years get all kinds of basic facts wrong. They may, for example, misidentify who has the appropriate jurisdiction to solve a problem, which means that even if the petition convinces the decision-makers reading it that something should be done, they don't have any power to actually do it. One obvious instance of this blunder that came across my Twitter feed was a petition asking the United Nations Secretary-General to change Florida fishing regulations. That, as you'll know from reading

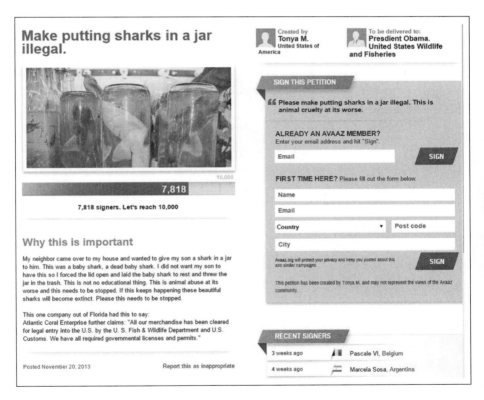

Make putting sharks in a jar illegal.

Created by
Tonya M.
United States of America

To be delivered to:
Presdient Obama.
United States Wildlife and Fisheries

SIGN THIS PETITION

❝❝ Please make putting sharks in a jar illegal. This is animal cruelty at its worse.

ALREADY AN AVAAZ MEMBER?
Enter your email address and hit "Sign".

Email   **SIGN**

FIRST TIME HERE? Please fill out the form below.

Name

Email

Country   ▼   Post code

City

Avaaz.org will protect your privacy and keep you posted about this and similar campaigns.   **SIGN**

This petition has been created by Tonya M. and may not represent the views of the Avaaz community.

10,000

7,818

7,818 signers. Let's reach 10,000

Why this is important

My neighbor came over to my house and wanted to give my son a shark in a jar to him. This was a baby shark, a dead baby shark. I did not want my son to have this so I forced the lid open and laid the baby shark to rest and threw the jar in the trash. This is not no educational thing. This is animal abuse at its worse and this needs to be stopped. If this keeps happening these beautiful sharks will become extinct. Please this needs to be stopped.

This one company out of Florida had this to say:
Atlantic Coral Enterprise further claims: "All our merchandise has been cleared for legal entry into the U.S. by the U. S. Fish & Wildlife Department and U.S. Customs. We have all required governmental licenses and permits."

Posted November 20, 2013   Report this as inappropriate

RECENT SIGNERS

3 weeks ago   Pascale VI, Belgium

4 weeks ago   Marcela Sosa, Argentina

A screenshot of a viral 2013 petition to ban the sale of sharks in jars, a Florida tourist curio. No sharks ever have been killed for this purpose, and this is not a serious conservation issue. *Courtesy of the author* this book, is just not how any of this works. Other petitions don't have a specific ask at all; they just say "We need to save the ocean" or "Something must be done." Asking someone to do something, but not saying what it is you want them to do, is just about the biggest waste of policy advocacy that I can imagine.

Other absurd petitions demand an end to things that are just not real problems. From 2011 to 2017, while I was earning my PhD in Florida, I saw more than a dozen online petitions asking the state to ban shark finning. Florida had already banned shark finning way back in 1994. Some of these petitions got 30,000 signatures or more, which is an unbelievable amount of effort and energy going toward a goal that was successfully achieved 25 years ago. And no, this is not a positive outcome with respect to "raising awareness," because it spreads wrong information. This is an example of actively contributing to public misunderstanding in a way that makes real solutions harder to enact. When scientists and environmentalists have to waste time debunking this nonsense, it means that we have less time available to work on real solutions to real problems.

Additionally, this kind of misinformation in viral petitions dilutes the limited resources of the conservation advocacy community. There are many people who want to help sharks, but we all have a limited amount of free time. When I saw many of my Facebook friends signing these petitions, I reached out to say "It's great that you want to help sharks. Can you please consider submitting a formal public comment on this important matter?" I lost count of the number of people who replied with some version of "Sorry, I've already done my part to help sharks this month" when it passed 100.

Submitting formal public comments in support of or in opposition to proposed management changes is an important action that you can take to help sharks. Signing and sharing an amateur-made online petition is extremely unlikely to help anything, and may cause harm in a lot of different ways. And if I may make one more plea here, if you don't have explicit training or experience with conservation advocacy, please, please, please don't write your own online petitions.

Volunteering/Donating to Help a Reputable Conservation Nonprofit (and NOT Starting Your Own Nonprofit)

Thankfully, when people hear about the problems facing the ocean, many feel inspired to do something to help. However, several times each month for the past decade, I've been approached by concerned members of the community who want to found their own ocean conservation nonprofits so they can help. That's potentially not great. As I wrote in my April 2020 Ask a Marine Biologist column in *Sport Diver* magazine, "If you care about sharks and want to help them, but don't have any training, experience, credentials, or knowledge relevant to policy change or conservation advocacy, I strongly recommend that you volunteer to help with an existing conservation nonprofit rather than start your own."

Let me be totally clear: if you have relevant knowledge, skills, or experience related to any aspect of environmental advocacy and policy change, there's a lot you can bring to the table here. But far too many people, realizing that conservation nonprofits exist and wanting to do *something*, seem to have decided creating their own nonprofit is the way to go.

I always ask folks inquiring about this possibility, "What will your new nonprofit do that's different from the many nonprofits that already exist?" Sometimes they have a great answer. They may have a new approach, a new angle, even an entirely innovative evidence-based policy solution. However, most of the time, the response I receive is something along the lines of "Huh, I didn't know there were any other shark conservation organizations." This tells me that they haven't really looked into what's going on, which makes it a lot less likely that their new organization is going to be able to do much to help.

Perhaps these questions are inevitable consequences of human nature. Some people would simply rather be President of the South Hills of Pittsburgh Save the Sharks Club than one of hundreds of volunteers supporting an existing effective organization. Maybe this kind of query is an innocent mistake caused by not doing enough research about what the ocean really needs and what organizations already exist. Whatever the cause, whatever the explanation, the desire to create a new organization can absolutely cause harm. It can dilute the donor pool, taking funds away from organizations that can (and do) make a huge impact, ones that urgently need help. It can also divert other important resources like public attention. And like useless online petitions, these organizations can unknowingly spread misinformation that makes it harder for ocean conservation professionals to do their jobs.

You learned in this book that there are a lot of great shark conservation nonprofits out there; you would do much more to assist sharks by just helping those nonprofits instead. One of the most effective things you can do is donate money or arrange a fundraising event. Sometimes these organizations need you to donate your time. Some have formal long-term volunteer positions, or education ambassadors who they train to help to raise awareness of effective solutions that need more public support. But realistically, the thing they're most likely to need from you is money, and just donating money to an effective nonprofit is far more likely to help save the ocean than creating your own new organization from scratch.

Following and Amplifying Experts (and NOT Sharing Nonsense from Fraudsters)

> Never before have so many people had so much access to so much knowledge and yet been so resistant to learning anything.
>
> —Tom Nichols, *The Death of Expertise: The Campaign Against Established Knowledge and Why it Matters*

If you ever heard of me before you bought this book, it was likely through my social media presence. Since 2009, I've been using Twitter to share expert information about shark science and conservation with a large and growing audience. The medium obviously has lots of downsides, but social media tools have made it easier than ever before in human history for experts to share their knowledge with the interested public. In a very real sense, it's never been easier for one person (or a small group of people) to change the world!

From a science-based conservation policy standpoint, the power of social media is clear. Lots of people want to support ocean conservation but don't necessarily know how to get the most bang for their buck, experts know how to effectively help but need more peoples' support, and social media can connect these two groups nicely. If you follow the right accounts, you'll be well informed about what's going on and how you can contribute to solutions. It's no accident that social media channels are widely used by ocean conservation professionals. This is a win-win, right? Unfortunately, it's just not that simple.

During my professional development workshops for scientists who want to learn about using Twitter to engage with the public, I share some tips and tricks. Here's one that I call David's First Law of the Internet: The wonderful thing about social media is that it gives everyone a voice, but the terrible thing about social media is that it gives *everyone* a voice. Just as social media makes it easy for experts to share facts with concerned members of the public, social media also makes it incredibly easy to spread misinformation.

As a scholar of the human dimensions of ocean conservation issues, I've looked at how misinformation related to sharks is spread online, results that I've shared throughout this book. A series of confusing and frustrating social media interactions inspired my postdoctoral fellowship research, which focused on understanding where people who aren't scientists learn about shark conservation issues and how accurate the information shared through those information pathways is.

My longtime social media followers have seen some . . . memorable conversations over the last decade between me and passionate but misinformed shark enthusiasts. I think my all-time favorite exchange occurred when my PhD lab at the University of Miami won a community education grant from Wells Fargo, and someone on Twitter became furious at me for "selling sharks to a bank." What exactly they believed the bank wanted with sharks was never made clear to me, but at least it was a great opportunity to make a bad pun about loan sharks. I get angrily told to "do my research" about shark conservation several times a month by someone whose only knowledge of the issue comes from incorrect viral information on social media. And a certain subset of ocean enthusiast social media also has an unfortunate tendency to blame every piece of bad news about marine life on the Fukushima nuclear disaster. As it turns out, we've known that sharks get cancer since the 1850s; it's not a new phenomenon caused by recent nuclear radiation.

Although these occasional one-off interactions can be entertaining, my time engaging with the public on social media has revealed widespread public misunderstanding about what's going on with shark conservation. One frustratingly persistent example exposes a concerning disconnect between experts and enthusiastic non-experts about the basic goal of shark conservation. As you've read, fully 90% of the shark scientists and management experts I've surveyed support the idea of sustainable shark fisheries whenever possible, as do 78% of the environmental nonprofit organization advocates whom I surveyed. From this, you might conclude that sustainable fisheries for sharks are a mainstream, commonly accepted idea—which is certainly true among experts. But when I've shared statements in support of sustainable shark fisheries on Twitter, I've gotten insults, harassment, and even death threats from people who believe banning all shark fishing everywhere (or sometimes banning all fishing of any kind) is obviously the only possible solution.

At one point, someone submitted a formal rebuttal to one of my scientific papers about sustainable shark fisheries to the same scientific journal that published my piece. This rebuttal claimed that not only was *I personally* corrupted by industry and therefore not to be trusted, but that *the entire field of fisheries science* was irretrievably compromised and all shark researchers were untrustworthy. Any scientists who support fishing, the author reasoned, are obviously evil or on the take. The rebuttal claimed that I'd been bribed by both the fishing industry and Asian organized crime families, and insisted that a criminal investigation should be opened. The proposed solution in this rebuttal (which was fortunately heavily edited before it was published, but is still available on the author's personal blog as of this writing) was not only for me to retract my paper but for me to be banned from publishing anything in the future.

Generally, successful environmental campaigns require working with people who agree with you on the issue of the day even if they don't agree with you on much else. Things become much more fraught, though, when you encounter a genuinely malicious opponent. I think threatening the life of someone you disagree with or threatening their family crosses a line, as is claiming that someone is involved in international criminal conspiracies and should be professionally sanctioned. Occasionally heated discussions with people I like and respect but disagree with have made me a better scientist, but the notion that people should agree to disagree should mean "I value my continued ability to work with you in the future more than I care about coming to an agreement about the resolution to this particular issue." This is just not always the case.

The observation that experts overwhelmingly support sustainable shark fisheries, yet enthusiastic non-experts on social media seem to have no idea that sustainable fishing is even a possibility, inspired my postdoctoral research focus. My team sought to determine why so many people who love sharks and want to save them seem to have no idea what huge majorities of experts believe the correct solutions to shark-related threats to be. Our analysis of almost 2,000 newspaper articles from around the world found that popular press coverage of sharks focuses overwhelmingly on one type of solution (bans on fishing for sharks or trade in shark products) at the expense of informing readers about other, more proven solutions (like sustainable fisheries management tools). In fact, the entire topic of sustainable fisheries management tools gets less media coverage than one small nonprofit that stages flashy protests directed at beloved brands like Starbucks that have very little connection to shark issues.

You may be wondering what Starbucks has to do with the global shark fin trade. I really tried to get a straight answer out of some of the organizers of this protest against "Shark Fin Bucks," but they largely seemed confused as to why I wasn't just supporting them since they claimed they were trying to save sharks. As near as I can tell from the group's promotional materials, the food importer/exporter that distributes Starbucks coffee in Asia is owned by another company. This other company also owns a small restaurant chain that serves shark fin soup. I personally don't think it's especially fair or useful to blame Starbucks for this tenuous connection, but flashy protests like the one targeting them due to this tenuous connection get more media coverage that all sustainable shark fisheries management policy tools combined. No wonder some people are a little confused as to what's really going on and how they can help.

As of this writing, social media companies have largely chosen not to intervene in slowing the spread of misinformation on their platforms, with the notable exception of conspiracy theories associated with the 2020 election and the COVID-19 pandemic. Currently, all we can do is ensure that as much accurate information as possible is shared, and that false information isn't allowed to stand unchallenged. Why is this helpful? Well, even people with naturally skeptical minds may think, "That doesn't sound quite right, but maybe there's something to it." In other words, even when they identify misinformation as being unreliable, they wrongly believe there could be a kernel of truth to it. Challenging such misinformation clearly and forcefully is therefore a critically important task for experts. You may not change the mind of the person you're arguing with, but you'll almost certainly change the minds of some people watching the exchange, even if you never interact with them directly.

Overall, though, my positive experiences on Twitter have outweighed my negative experiences in both frequency and magnitude. I have more positive experiences than I have negative experiences, and the positive ones are more positive than the negative ones are negative. I certainly understand social media isn't for everyone, but my time on Twitter has made me a better writer, a better scientist, and a better teacher. I've also really enjoyed helping to build an online community of experts.

Following credible experts and sharing their calls for assistance is a great thing you can do to help, but please take care not to contribute to the spread of misinformation. If you're not sure if something's true and you're not sure if the person saying it is trustworthy, it probably isn't and they probably aren't. A list of some reliable expert shark conservation voices you can follow on social media is available on this book's website.

Diving with Sharks, but NOT Swimming with Dangerous Fools

I can't believe that "please don't grab the 18-foot-long wild predator" is something that needs to be explicitly said out loud.

—My comment in a 2019 *Washington Post* article on the rise of riding sharks for Instagram likes

Wildlife tourism is one of the most popular and visible types of potential conservation action, but are you truly contributing to the cause of protecting sharks by participating in these activities? The short if less than satisfying answer is it depends. Some wildlife tourism can help some sharks sometimes, but it is not the ultimate solution to every shark conservation crisis, as some people claim. The argument for the effectiveness of wildlife tourism is straightforward even if the evidence for it is mixed: if fishermen kill a shark, they can sell it once, but if that same shark is alive on a reef it can attract tourists to scuba dive with it for years, thus making sharks "worth more alive than dead." Yes, it is true that wildlife tourism can contribute by providing locals with a financial incentive to protect sharks instead of fishing for and killing them. However, the kind of emotional appeal that animal encounters typically generate doesn't work in every context. Wildlife tourists can't encounter all threatened shark species; some species are highly migratory, while others live in the open ocean or the deep sea, far from where scuba divers can visit (see "Meet a Scientist: The Limits of Wildlife Tourism and a Safer Path for Shark Science Education" in chapter 8 for more information). Of the most common shark species encountered by wildlife tourism operations, most are not especially threatened. This means that, although wildlife tourism contributes to local shark conservation efforts in some places, it is not a major solution to global-scale threats faced by many of the most vulnerable species.

Additionally, it's important to track whether the people losing income when new conservation regulations are enacted, like local fishers, are the same people profiting from a rise in wildlife tourism. In the Bahamas, for example, lots of divers seeking sharktastic adventures spend their entire vacations on liveaboard dive vessels owned by people from Florida. But Bahamian fishers, the people most financially impacted by stricter regulations on fisheries, may not ever see a cent of profit from these businesses.

If you're choosing to vote with your wallet, then which scuba operator you choose matters. There are great conservation-minded scuba operators who work in ways that are safe for sharks and humans. They take care to educate their customers with real facts about shark conservation and how people can help. Then there are the aforementioned "macho

This person is not helping save sharks. He has flipped a shark over, I assume, because he thinks it makes for a cool photo. Recall, though, that tonic immobility causes physiological stress to sharks. *Courtesy of Mozcashew1, Wikimedia Commons*

cowboy idiots," who not only fail to share factual information about what the threats to sharks are and how people can help, but too often engage in harmful behavior like grabbing, riding, hugging, or kissing sharks. This is wildlife harassment, not outreach. Suffice it to say that you should not hug or kiss large wild predators—and while I have your attention, don't run with scissors and look both ways when you cross the street. Can you imagine a land-based wildlife tour where people knelt by a stream and tried to hand-feed grizzly bears while climbing on top of them?

When I criticize macho cowboy idiot behavior, it leads to some common types of pushback. Here are the kinds of excuses I hear, along with my responses to each statement:

"That's not wildlife harassment because I don't personally think it causes any harm."

Sorry, friends, grabbing a wild animal for no good reason is indeed a textbook case of wildlife harassment; it may even be a prosecutable offense in some jurisdictions. And like a great many issues, whether it causes harm is an objective measurable fact that isn't influenced by your opinion. Experts have studied the impacts of having natural behavior disrupted extensively and found that it causes physiological stress. Harassing behaviors may scare a shark away from a feeding opportunity, causing them to spend precious energy on something other than eating, swimming, and making baby sharks.

"Getting physical with sharks is good, actually, because it raises awareness."

Does it though? Awareness of what? I've never really understood this argument. It seems to posit that watching someone annoy a wild animal and not get killed by it will somehow cause people to want to save that animal. Also, the idea that sharks are dangerous is not the biggest, let alone the only threat they face, so even if this kind of wildlife harassment did help debunk that myth, it wouldn't be that useful.

"The people who do this are brave conservation heroes. Why are you criticizing them?"

Putting yourself in unnecessary danger is not brave. It's, well, unnecessary. I have lots of conservation heroes—you've met several of them in this book. One thing they all have in common is that they've actually engaged in some manner of conservation. Wildlife harassment is not conservation. And no one is above criticism.

"Aren't we all on the same team?"

I proudly work alongside many people who have lots of different environmental ideologies in support of shared goals. Even when we disagree on 90% of the issues I'm happy to work together on that 10% where we agree. The people who engage in wildlife harassment for Instagram likes are not on our team. They are not involved in serious approaches to ocean conservation, and do not respond positively to constructive suggestions for improvement (to put it mildly).

"Even if this behavior causes harm, surely it's better for sharks than shark finning is?"

This is an argument that might make sense if the only two possible ways humans could interact with sharks involved wastefully slaughtering them or kissing them. There are, happily, many other options.

While hiring an eco-friendly dive operator to go swimming with sharks can potentially contribute to shark conservation efforts by convincing locals to keep healthy shark populations around instead of overfishing them, swimming with sharks is not going to save most of the world's sharks. And unless you are thoughtful and informed when choosing an operator, you could be doing more harm than good.

Take Other Actions to Help the Ocean (Even If They Don't Necessarily Help Sharks Directly)

We can (and we should) do this for our sake, our children's sake, and the sake of all our fellow travelers on this miraculous ocean planet we call home.

—From Philippe Cousteau's foreword to David Helvarg's *50 Ways to Save the Ocean*

Assuming that you aren't currently eating endangered species of shark for breakfast, lunch, and dinner during breaks from your job bulldozing coral reefs, the above recommendations are probably the biggest things you can accomplish in your everyday life to help with shark conservation. However, I understand that the notion of waiting until there's an opportunity to submit a public comment, which may only happen twice a year, isn't very exciting when you're eager to pitch in. And if you're drowning in student loans it could be hard to find money to donate to reputable nonprofits. Here, then, are three big things you can do right away to help make oceans heathier for *all* sea creatures:

Purchase and eat sustainable seafood, and don't eat unsustainable seafood. We know that overfishing and bycatch represent the biggest threat to sharks and one of the biggest threats to the ocean overall. However, we are absolutely not at the point where the science indicates that the only solution is for everyone to give up seafood entirely. Instead, since these issues are caused by *unsustainable* fishing practices, the solution can involve promoting *sustainable* fishing practices. What makes seafood sustainable? There are several trains of thought here, but it's generally agreed upon that fishers engaged in sustainable practices (1) don't overfish—instead, they take care not to remove too many fish from the population; (2) create low levels of harmful bycatch, especially of endangered species; (3) cause limited ecosystem-wide impacts with their boats and gear; and (4) adhere to clear, enforceable rules that make sure all this happens. A variety of resources for locating sustainable seafood options are provided on this book's website.

Use less single-use plastic. Although plastic pollution isn't a big threat to sharks, it is very dangerous for many species of marine life, including those that are essential to sharks' habitats. We should do all we can to reduce our plastic consumption. This includes both systemic change, especially at the manufacturing level, and changes in our personal consumption habits. If you're able (and as we've discussed, not everyone is), try not to use plastic straws or plastic cutlery. Take care to properly dispose of packaging. Try to rely on reusable water bottles and coffee cups rather than single-use plastic options. Take cloth shopping bags to the grocery store so that you don't have to use plastic grocery bags. And if you're near a beach, you can participate in an organized beach cleanup.

Reduce your carbon footprint. I'm sure you've heard plenty about climate change as a threat to the ocean (and the rest of the planet) through sources other than this book. I won't belabor those points here other than to remind you that climate change and ocean acidification are major environmental issues, even if they don't happen to directly affect sharks very much. But we've seen how the health of one part of the ecosystem can impact other parts, and anything we can do to reduce CO_2 emissions, including but not limited to reducing our personal carbon footprints, helps the ocean as a whole. (Yes, the concept of a carbon footprint was literally invented by energy companies to make you think about the harm *you* cause and not the harm *they* cause, but there's still value in working to minimize harm on the individual level.)

A Quick Note on "Doing Something"

At least they're doing something to help; all you do is complain!

—The gist of approximately eleventy billion angry tweets I receive when I criticize a harmful amateur conservation initiative

Human beings have a bias toward action, something that seems especially true when it comes to the Kids These Days™ of Generation Z.

When people hear about a problem, our instinct is to look for ways to fix it. However, as I hope this chapter has made clear, our methods matter. And that's why I despise the phrase "At least they're doing something" as an argument. "Doing something" is not inherently good. Doing something that makes the problem worse is objectively worse than doing nothing. Doing something that doesn't help solve the problem while taking resources away from effective solutions is just as bad. The ocean faces many threats, and these threats need to be tackled. However, instead of just diving into the first idea that you come up with, please consider listening to experts about what needs doing, and what is just wasting time or causing harm.

The ocean needs your help, and so do the many people working to save it. Your time and effort can be incredibly valuable; please make sure you contribute in the most effective ways you can.

Afterword

Don't love the ocean too much. It won't love you back.

—Dick Richie (Mark Hengst), *Mega Shark Versus Giant Octopus*

One of my goals in writing this book was to share my delight and wonder about these amazing animals that so many people know only as a menace to fear when they go swimming. Another was to convince people that not only are sharks are worth saving, but that we need to act now to save them.

My last (and trickiest) goal in writing this book was to help people who already love sharks and want to save them, but have been misinformed about key facts that impact shark conservation threats and solutions. The ways that I see shark conservation discussed by concerned members of the public are often wildly different from the conversations among scientists, natural resource managers, and professional environmental advocates. I hope this book has helped to bridge the gap between shark fans and shark scientists a little and demonstrated why the problems we face are more complicated than can be conveyed with simple slogans. More people want to help sharks than ever before, but how we try to help them matters.

While I've tried to provide a comprehensive introduction to the world of shark conservation science and policy, this field is too vast to be contained in a single book. Therefore, I'd like to leave you with some resources to help you find more information and continue your education. There's a lot of inaccurate information about shark conservation on the internet, so first, let me point you again to the book's website, which can be found at https://jhupbooks.press.jhu.edu/title/why-sharks-matter. It features many carefully vetted links and resources that you can use to learn more about all the topics I discussed in this book from trustworthy, reliable sources.

Many of the experts featured in this book from the world of shark science and environmental advocacy were chosen explicitly because they share information for the interested public on social media; you can find their social media handles on the book's website. You can also follow me on social media (Twitter, Facebook, and Instagram @WhySharks-Matter), where I'm always happy to answer any questions anyone has about sharks. There, I share new updates from the science and policy world, along with messages from my colleagues. I also alert followers when there are proposed conservation regulations they can comment on. Keep in mind that I'm happy to speak to your science class, local environmental club, or community organization.

The introduction to this book included my favorite quotation about conservation, which is by Senegalese conservationist Baba Dioum: "In the end we will conserve only what we love; we will love only what we understand; and we will understand only what we are taught." I hope that I've taught you some interesting things about sharks, and that if you still don't love them as much as I do, now you at least understand them enough to recognize that sharks matter and that they need our help.

Thanks for taking this journey with me!

David Shiffman, PhD
Washington, DC

ACKNOWLEDGMENTS

I've been working on this book in various forms for well over a decade. (I actually envisioned *Why Sharks Matter* as a book title well before I used it to create my social media handles back in 2009.) This long journey means I have an awful lot of people to thank.

First and foremost, I want to express my gratitude to my family, who have always encouraged me to pursue my unusual career goals. They've shown me how valuable it is to never stop learning and how important it is to appreciate the wonders of the natural world. My parents and my brother have also been invaluable copyeditors for various drafts of this book and for countless articles I cite within it. My accomplishments would not have been possible without them all standing by me. Dad has also set the record for most books pre-ordered, so if you're one of his golf buddies and you get this book as a Christmas present, I hope you enjoy it and that you receive the customary nice bottle of liquor next year.

I also want to thank my many teachers and mentors. As this book has hopefully made clear, the world of science-based conservation can be technical and complicated. I'm grateful to those who have taught me so much about how to understand sharks, how to be a successful scientist, how to devote my work to protecting threatened species, and how to explain that work to others. I especially want to express appreciation to my teachers, whose tolerance and vision allowed me to push my studies outside the boundaries of their normal research. Thank you to my undergraduate advisor, Dan Rittschoff, for agreeing to let me do an undergraduate project focusing on stingrays instead of the blue crabs he's best known for studying; my master's advisor, Gorka Sancho, for giving me the flexibility to design my own research project outside bony fish ecology (and for introducing me to #BestShark); and my PhD advisor, Neil Hammerschlag, for giving me the freedom to pursue the side projects and side hustles that have become the focus of my work. My postdoctoral supervisor, Nick Dulvy, introduced me to the world of global-scale science-based conservation and showed me the power of collaboration, and my postdoctoral supervisor, Lara Ferry, has demonstrated by her excellent example how to be an effective educator and a supportive mentor. Sonja Fordham saw the value of my online outreach and took the time to teach me all about policy nuance when I was a master's student. I still learn something new every time we chat. I also wouldn't be where I am today without lessons I learned from many other teachers and mentors, including Alec Motten, Julie Reynolds, Grace Upshaw, Craig Plante, Dave Owens, Erik Sotka, Tony Harold, Bryan Frazier, Dan Abel, Dean Grubbs, Dave Letson, Michael Owens, Gina Maranto, and Kenny Broad.

I'm incredibly grateful for my friends and colleagues from the world of marine biology, ocean conservation, and science communication, who helped transform me into an interdisciplinary scientist. You've met 11 of my brilliant and inspirational colleagues in this book's "Meet a Scientist / Meet a Conservationist" features; in addition to them, I want to thank some of the many others who have taught me so much, including Andrew Thaler (who claims I was already talking about writing a version of this book when we met in 2003), Amy Freitag, Julia Wester, Chris Mull, Jenny Bigman, Lindsay Davidson, Colin Simpfendorfer, Pete Kyne, Lisa Whitenack, John Carlson, Tobey Curtis, Karyl Brewster-Geisz, Bob Hueter, Isabelle Cote, Jeremy Vaudo, Sora Kim, Toby Daly-Engel, Steve Cooke, Duncan Irschick, Jeff Carrier, Misty Paig-Tran, Sarah Fowler, Chris Pepin-Neff, Jim Gelsleichter, Kara Yopak, James Sulikowski, Beth Polidoro, Jayne Gardner, Hannah Medd, Katie Matthews, Jim Wharton, Mariah Pfleger, Angelo Villagomez, Rick MacPherson, Michelle Heupel, Yannis Papastamatiou, Matt Ajemian, Chris Bird, Al Dove, Les Kaufman, Craig McClain, Miriam Goldstein, Rebecca Helm, Solomon David, Tricia Meredith, Josh Drew, Emily Darling, Stacy Farina, Amani Webber-Schultz, Chris Bedore, Derek Burkholder, Angela Bednarek, Rachel Pendergrass, Emily Knight, Luiz

Rocha, Asha De Vos, Dave Kerstetter, Edd Hind-Ozan, Chip Cotton, Jake Jerome, Christian Pankow, Nick Perni, Vicky Vasquez, Scott Wallace, Tonya Wiley, Gene Helfman, Kristine Stump, Ian Campbell, and Jen Wyffels. I'm a better scientist, a better teacher, and a better writer because y'all are in my life, and I always come away from our chats (and/or arguments) edified and inspired. Anything interesting in this book I learned from one of these folks, and any errors are entirely mine.

I want to especially thank the Liber Ero Postdoctoral Fellowship and my colleagues there, particularly Brett, Aerin, SJ, Cody, Karen, Adam, Christina, Sheila, Liz, Jen, Brynn, Jean, Richard, Diane, Nathan, and our fearless leader, Sally Otto, for making me a better conservation scientist and encouraging me to use my time as a fellow to start writing this book in earnest.

Thanks also to my ScienceOnline friends, especially former director Karyn Murphy, for introducing me to the power of social media for communicating science. It was at a book writing workshop at ScienceOnline 2010 that I first wrote the outline for what became a big chunk of this book!

I specifically want to recognize that some individuals mentioned here generously dedicated time and energy to fact-checking this book. Nick Dulvy, Sonja Fordham, Ali Hood, Tobey Curtis, Kristian Parton, and Chris Pepin-Neff checked big chunks of text, and Catherine Macdonald and Chuck Bangley read the whole thing. Thanks to you all for helping make the book better!

Thanks to the many editors who have commissioned me to write the popular press articles I've referenced throughout this book, giving me the chance to hone my craft as a writer for general audiences. I especially want to thank Laura Helmuth, Mike Lemonick, Dean Visser, Rachael Bale, Katie Burke, Colin Schultz, Adrienne Mason, Jude Isabella, John Platt, Dave Carriere, Becca Hurley, and Allie Gillespie.

Speaking of editors, thanks to the great folks at Johns Hopkins University Press, especially my acquisitions editor, Tiffany Gasbarrini, for approaching me to write this book. All of her hard work made my ideas presentable! Thanks also to Hilary Jacqmin and Esther Rodriguez for their editorial efforts. And a special thank-you to my publicity manager,

Kathryn Marguy, for her dedication to getting my books into the hands of readers.

Finally, thanks to all of you out there for reading my words, whether in this book, in my scientific or popular press articles, or on social media. Remember to please direct all feedback, including but hopefully not limited to hate mail, to WhySharksMatter@gmail.com.

BIBLIOGRAPHY

Introduction
Shiffman, D. S., Frazier, B. S., Kucklick, J. R., Abel, D., Brandes, J., & Sancho, G. (2014). Feeding ecology of the sandbar shark in South Carolina estuaries revealed through δ13C and δ15N stable isotope analysis. *Marine and Coastal Fisheries, 6*(1), 156–169. https://www.tandfonline.com/doi/abs/10.1080/1942 5120.2014.920742

Chapter 1: Shark Basics, and Fun Facts to Keep You Reading
Chapman, D. D., Shivji, M. S., Louis, E., Sommer, J., Fletcher, H., & Prodöhl, P. A. (2007). Virgin birth in a hammerhead shark. *Biology Letters, 3*(4), 425–427. https://royalsocietypublishing.org/doi/abs/10.1098/rsbl.2007.0189
Joung, S. J., Chen, C. T., Clark, E., Uchida, S., & Huang, W. Y. (1996). The whale shark, Rhincodon typus, is a livebearer: 300 embryos found in one "megamamma" supreme. *Environmental Biology of Fishes, 46*(3), 219–223. https://link.springer.com/article/10.1007/BF00004997
Keller, B. A., Putman, N. F., Grubbs, R. D., Portnoy, D. S., & Murphy, T. P. (2021). Map-like use of Earth's magnetic field in sharks. *Current Biology.* https://www.sciencedirect.com/science/article/abs/pii/S0960982221004760
Macdonald, C. (2020, October 19). Shark biology and fisheries. *Save Our Seas Magazine.* https://saveourseas.com/update/shark-biology-and-fisheries/
McClain, C. (2017, December 31). How do we know megalodon doesn't still exist? *Deep Sea News.* https://www.deepseanews.com/2017/12/how-we-know -megalodon-doesnt-still-exist/
McCoy, T. (2014, March 10). Stingray stabbed Crocodile Hunter Steve Irwin "hundreds of times," claims cameraman. *The Washington Post.* https://www .washingtonpost.com/news/morning-mix/wp/2014/03/10/stingray-stabbed -crocodile-hunter-steven-irwin-hundreds-of-times/
Nakaya, K., White, W. T., & Ho, H. C. (2020). Discovery of a new mode of oviparous reproduction in sharks and its evolutionary implications. *Scientific Reports, 10*(1), 1–12. https://www.nature.com/articles/s41598-020-68923-1
Outdoor Hub Reporters. (2015, June 9). Florida man attempts to sell shark outside supermarket. *Outdoor Hub News.* https://www.outdoorhub.com /news/2015/06/09/florida-man-attempts-sell-shark-outside-supermarket/

Price, W. (2018, July 19). Just in time for Shark Week, Florida scientists discover deepwater shark species named "Genie's dogfish." *USA Today.* https://www.usatoday.com/story/news/nation-now/2018/07/19/shark-week-scientists-discover-deepwater-shark-species-genies-dogfish/799366002/

Shiffman, D. S. (2019, June 27). Amazing shark facts you haven't heard a million times. *Scuba Diving Magazine.* https://www.scubadiving.com/uncommon-shark-facts

Shiffman, D. S. (2019, September 6). Do sharks have bones? Why you don't tend to see fossilized shark skeletons. *Sport Diver Magazine.* https://www.sportdiver.com/do-sharks-have-bones

Shiffman, D. S. (2019, September 12). Shark species you've probably never heard of, but desperately need your help. *Scuba Diving Magazine.* https://www.scubadiving.com/critically-endangered-shark-species

Shiffman, D. S. (2019, December 10). What's the deal with a goblin shark's crazy snout? *Sport Diver Magazine.* https://www.sportdiver.com/goblin-shark-snout

Shiffman, D. S. (2020, February 25). How far can sharks smell blood? *Sport Diver Magazine.* https://www.sportdiver.com/how-far-can-sharks-smell-blood

Shiffman, D. S. (2020, October 21). This mommy shark has a unique way of making babies. *Hakai Magazine.* https://www.hakaimagazine.com/news/this-mommy-shark-has-a-unique-way-of-making-babies/

Shiffman, D. S. (2020, November 6). Sharks never run out of teeth. *Scientific American.* https://www.scientificamerican.com/gallery/sharks-never-run-out-of-teeth/

Shiffman, D. S. (2021, March 10). Do sharks die if they stop swimming? *Sport Diver Magazine.* https://www.sportdiver.com/do-sharks-die-if-they-stop-swimming

Viegas, J. (2013, August 5). One female shark's litter may have many dads, study finds. *NBC News.* https://www.nbcnews.com/sciencemain/one-female-sharks-litter-may-have-many-dads-study-finds-6c10848675

Whittle, P. (2021, May 16). Sharks use Earth's magnetic field as a GPS. *Associated Press.* https://apnews.com/article/sharks-gps-magnetic-field-abf97cf60bb15f7fbf3bfed74671e398

Wolfe, S. (2003, Fall Issue). Who killed Jane Stanford? *Stanford Magazine.* https://stanfordmag.org/contents/who-killed-jane-stanford

Wu, K. (2020, October 13). Coronavirus vaccine makers are not mass-slaughtering sharks. *The New York Times.* https://www.nytimes.com/2020/10/13/science/sharks-vaccines-covid-squalene.html

Yong, E. (2013, July 10). Thresher sharks hunt with huge weaponised tails. *National Geographic.* https://www.nationalgeographic.com/science/article/thresher-sharks-hunt-with-huge-weaponised-tails

Chapter 2: Sharks Are Not a Threat to Humans

Anthony, S. (2008, August 12). Innovation lessons from Lisa Simpson's rock. *Harvard Business Review*. https://hbr.org/2008/08/innovation-lessons-from -lisas

Dehnart, A. (2020, August 4). Shark Week and Sharkfest's experts are mostly white men. Why? *Reality Blurred*. https://www.realityblurred.com /realitytv/2020/08/shark-week-shark-fest-2020-diversity/

Hazin, F. H., & Afonso, A. S. (2014). A green strategy for shark attack mitigation off Recife, Brazil. *Animal Conservation, 17*(4), 287–296. https:// zslpublications.onlinelibrary.wiley.com/doi/abs/10.1111/acv.12096

Jacques, P. J. (2010). The social oceanography of top oceanic predators and the decline of sharks: A call for a new field. *Progress in Oceanography, 86*(1–2), 192–203. https://www.sciencedirect.com/science/article/pii /S0079661110000352

Muter, B. A., Gore, M. L., Gledhill, K. S., Lamont, C., & Huveneers, C. (2013). Australian and US news media portrayal of sharks and their conservation. *Conservation Biology, 27*(1), 187–196. https://conbio.onlinelibrary.wiley.com /doi/abs/10.1111/j.1523-1739.2012.01952.x

Neff, C. (2015). The *Jaws* effect: How movie narratives are used to influence policy responses to shark bites in Western Australia. *Australian Journal of Political Science, 50*(1), 114–127. https://www.tandfonline.com/doi/abs/10.10 80/10361146.2014.989385

Neff, C., & Hueter, R. (2013). Science, policy, and the public discourse of shark "attack": a proposal for reclassifying human–shark interactions. *Journal of Environmental Studies and Sciences, 3*(1), 65-73. https://link.springer.com /article/10.1007/s13412-013-0107-2

Nelson, V. (2006, February 13). Peter Benchley: How *Jaws* author became a shark conservationist. *Los Angeles Times*. https://www.latimes.com/archives /la-xpm-2006-feb-13-me-benchley13-story.html

Nosal, A. P., Keenan, E. A., Hastings, P. A., & Gneezy, A. (2016). The effect of background music in shark documentaries on viewers' perceptions of sharks. *PLOS ONE, 11*(8), e0159279. https://journals.plos.org/plosone /article?id=10.1371/journal.pone.0159279

Shiffman, D. S. (2014). Keeping swimmers safe without killing sharks is a revolution in shark control. *Animal Conservation, 17*(4), 299–300. https:// zslpublications.onlinelibrary.wiley.com/doi/abs/10.1111/acv.12155

Shiffman, D. S. (2014, March 3). Shark riders pose threat to conservation gains made with diving ecotourism. *Scientific American*. https:// www.scientificamerican.com/article/shark-riders-pose-threat-to -conservation-gains-made-with-diving-ecotourism-slide-show1/

Shiffman, D. S. (2014, April 17). Cull kill includes small tiger sharks along with intended victims. *Scientific American*. https://www.scientificamerican.com /article/cull-kill-includes-small-tiger-sharks-along-with-intended-victims -video/

Shiffman, D. S. (2014, May 6). 24 species of sharks that have killed fewer people than Jack Bauer on 24. *Southern Fried Science*. https://www.southernfriedscience.com/24-species-of-sharks-that-have-killed-fewer-people-than-jack-bauer-on-24/

Shiffman, D. S. (2014, May 8). Do dolphins protect people from sharks? *Slate Magazine*. https://slate.com/technology/2014/05/adam-walker-saved-by-dolphins-from-shark-attack-the-myth-that-dolphins-protect-people.html

Shiffman, D. S. (2014, August 11). Shark Week lied to scientists to get them to appear in documentaries. *Gizmodo*. https://gizmodo.com/shark-week-lied-to-scientists-to-get-them-to-appear-in-1619280737

Shiffman, D. S. (2014, September 16). No, Jaws is not lurking off the Cornish coast: scaremongering claims could easily have been avoided. *New Scientist*. https://www.newscientist.com/article/dn26217-no-jaws-is-not-lurking-off-the-cornish-coast/

Shiffman, D. S. (2015, June 19). 40 years of bad science: How *Jaws* got everything wrong about sharks. *Gizmodo*. https://gizmodo.com/40-years-of-bad-science-how-jaws-got-everything-wrong-1712384448

Shiffman, D. S. (2015, July 6). Shark Week is upon us, and as a shark scientist, I both love and hate it. *Vox*. https://www.vox.com/2015/7/6/8886743/shark-week-2015-science

Shiffman, D. S. (2018, July 24). Shark scientists explain what's right and what's wrong with Shark Week. *The Washington Post*. https://www.washingtonpost.com/news/animalia/wp/2018/07/24/shark-scientists-explain-whats-right-and-whats-wrong-with-shark-week/

Shiffman, D. S. (2020, February 21). How to read the shark attack file report like a marine biologist. *Scuba Diving Magazine*. https://www.scubadiving.com/how-to-read-shark-attack-file-report-like-marine-biologist

Shiffman, D. S. (2020, August 27). How technology is making shark news more ridiculous. *Scuba Diving Magazine*. https://www.scubadiving.com/how-technology-is-making-shark-news-more-ridiculous

Thaler, A. D., & Shiffman, D. (2015). Fish tales: Combating fake science in popular media. *Ocean & Coastal Management, 115*, 88–91. https://www.sciencedirect.com/science/article/pii/S0964569115000903

Vásquez, V. E., Ebert, D. A., & Long, D. J. (2015). Etmopterus benchleyi n. sp., a new lanternshark (Squaliformes: Etmopteridae) from the central eastern Pacific Ocean. *Journal of the Ocean Science Foundation, 17*, 43–55.

Wetherbee, B. M., Lowe, C. G., & Crow, G. L. (1994). A review of shark control in Hawaii with recommendations for future research. *Pacific Science, 48*(2): 95–115.

Yuhas, A. (2021, July 20). Don't call them "shark attacks," scientists say. *The New York Times*. https://www.nytimes.com/2021/07/20/science/shark-attacks.html

Zachos, E. (2019, June 29). Why are we afraid of sharks? There's a scientific explanation. *National Geographic*. https://www.nationalgeographic

.co.uk/animals/2019/06/why-are-we-afraid-sharks-theres-scientific
-explanation#:~:text=They%20learned%20fear%20as%20an,re%20very%20
prone%20to%20fear.%22

Chapter 3: The Ecological Significance of Sharks

Atwood, T. B., Connolly, R. M., Ritchie, E. G., Lovelock, C. E., Heithaus,
M. R., Hays, G. C., Fourqurean, J. W., & Macreadie, P. I. (2015). Predators
help protect carbon stocks in blue carbon ecosystems. *Nature Climate Change*,
5(12), 1038–1045. https://www.nature.com/articles/nclimate2763

Burkepile, D. E., & Hay, M. E. (2007). Predator release of the gastropod
Cyphoma gibbosum increases predation on gorgonian corals. *Oecologia*,
154(1), 167–173. https://link.springer.com/article/10.1007/s00442-007
-0801-4

Chapman, D. D., Pikitch, E. K., & Babcock, E. A. (2006). Marine parks need
sharks? *Science*, *312*(5773), 526–528. https://science.sciencemag.org
/content/312/5773/526.4.abstract

Clementi, G. M., Bakker, J., Flowers, K. I., Postaire, B. D., Babcock, E. A.,
Bond, M. E., . . . & Chapman, D. D. (2021). Moray eels are more common
on coral reefs subject to higher human pressure in the greater Caribbean.
iScience, *24*(3), 102097. https://www.sciencedirect.com/science/article/pii
/S2589004221000651

Desbiens, A. A., Roff, G., Robbins, W. D., Taylor, B. M., Castro-Sanguino, C.,
Dempsey, A., & Mumby, P. J. (2021). Revisiting the paradigm of shark-driven
trophic cascades in coral reef ecosystems. *Ecology*, *102*(4), e03303. https://
esajournals.onlinelibrary.wiley.com/doi/full/10.1002/ecy.3303

Dicken, M. L., Hussey, N. E., Christiansen, H. M., Smale, M. J., Nkabi, N.,
Cliff, G., & Wintner, S. P. (2017). Diet and trophic ecology of the tiger
shark (Galeocerdo cuvier) from South African waters. *PLOS ONE*, *12*(6),
e0177897. https://journals.plos.org/plosone/article?id=10.1371/journal.
pone.0177897

Drymon, J. M., Feldheim, K., Fournier, A. M. V., Seubert, E. A., Jefferson, A. E.,
Kroetz, A. M., & Powers, S. P. (2019). Tiger sharks eat songbirds: Scavenging
a windfall of nutrients from the sky. *Ecology*, (9), e02728. https://esajournals
.onlinelibrary.wiley.com/doi/abs/10.1002/ecy.2728

Dulvy, N. K., Freckleton, R. P., & Polunin, N. V. (2004). Coral reef cascades and
the indirect effects of predator removal by exploitation. *Ecology Letters*, *7*(5),
410–416. https://onlinelibrary.wiley.com/doi/abs/10.1111/j.1461
-0248.2004.00593.x

Estes, J. A., Terborgh, J., Brashares, J. S., Power, M. E., Berger, J., Bond, W. J.,
. . . & Wardle, D. A. (2011). Trophic downgrading of planet Earth. *Science*,
333(6040), 301–306. https://science.sciencemag.org/content/333/6040/301
.abstract

Estes, J. A., Tinker, M. T., Williams, T. M., & Doak, D. F. (1998). Killer whale
predation on sea otters linking oceanic and nearshore ecosystems. *Science*,

282(5388), 473–476. https://science.sciencemag.org/content/282/5388 /473.abstract

Evans, G. (2020, July 12). Sorry everyone but that giant bird in the viral video wasn't carrying a shark. *Indy100*. https://www.indy100.com/offbeat/shark -myrtle-beach-spanish-mackerel-video-9614686

Frid, A., Baker, G. G., & Dill, L. M. (2008). Do shark declines create fear -released systems? *Oikos*, *117*(2), 191–201. https://onlinelibrary.wiley.com /doi/abs/10.1111/j.2007.0030-1299.16134.x

Gander, K. (2020, November 11). Killer whales who rip open great white sharks, eat their livers, may explain disappearance. *Newsweek*. https://www.newsweek .com/killer-whales-great-white-sharks-eat-livers-predators-disappear-south -africa-1548999

Grubbs, R. D., Carlson, J. K., Romine, J. G., Curtis, T. H., McElroy, W. D., McCandless, C. T., Cotton, C. F., & Musick, J. A. (2016). Critical assessment and ramifications of a purported marine trophic cascade. *Scientific Reports*, *6*(1), 1–12. https://www.nature.com/articles/srep20970

Hammerschlag, N., Luo, J., Irschick, D. J., & Ault, J. S. (2012). A comparison of spatial and movement patterns between sympatric predators: Bull sharks (Carcharhinus leucas) and Atlantic tarpon (Megalops atlanticus). *PLOS ONE*, *7*(9), e45958. https://journals.plos.org/plosone/article?id=10.1371/journal .pone.0045958& xid=17259,15700002,15700021,15700186,15700191, 15700256,15700259,15700262

Hammerschlag, N., Williams, L., Fallows, M., & Fallows, C. (2019). Disappearance of white sharks leads to the novel emergence of an allopatric apex predator, the sevengill shark. *Scientific Reports*, *9*(1), 1–6. https:// www.nature.com/articles/s41598-018-37576-6

Heithaus, M. R., & Dill, L. M. (2002). Food availability and tiger shark predation risk influence bottlenose dolphin habitat use. *Ecology*, 83(2), 480–491. https://esajournals.onlinelibrary.wiley.com/doi/abs/10.1890/0012 -9658(2002)083[0480:FAATSP]2.0.CO;2

Leigh, S. C., Papastamatiou, Y. P., & German, D. P. (2018). Seagrass digestion by a notorious "carnivore." *Proceedings of the Royal Society B: Biological Sciences*, *285*(1886), 20181583. https://royalsocietypublishing.org/doi/abs/10.1098 /rspb.2018.1583

Lloyd, J. (2017, June 14). The Chesapeake Bay's misguided war on the ray. *Hakai Magazine*. https://www.hakaimagazine.com/news/chesapeake-bays-misguided -war-ray/

Madin, E. M., Madin, J. S., & Booth, D. J. (2011). Landscape of fear visible from space. *Scientific Reports*, *1*(1), 1–4. https://www.nature.com/articles /srep00014

Myers, R. A., Baum, J. K., Shepherd, T. D., Powers, S. P., & Peterson, C. H. (2007). Cascading effects of the loss of apex predatory sharks from a coastal ocean. *Science*, *315*(5820), 1846–1850. https://science.sciencemag.org /content/315/5820/1846.abstract

Roff, G., Doropoulos, C., Rogers, A., Bozec, Y.-M., Krueck, N., Aurellado, E., Priest, M., Birrell, C., & Mumby, P. J. (2016). Reassessing shark-driven trophic cascades on coral reefs: A reply to Ruppert et al. *Trends in Ecology & Evolution, 31*(8), 587–589.

Ruppert, J. L., Travers, M. J., Smith, L. L., Fortin, M. J., & Meekan, M. G. (2013). Caught in the middle: combined impacts of shark removal and coral loss on the fish communities of coral reefs. *PLOS ONE, 8*(9), e74648. https://journals.plos.org/plosone/article?id=10.1371/journal.pone.0074648

Sherman, C. S., Heupel, M. R., Moore, S. K., Chin, A., & Simpfendorfer, C. A. (2020). When sharks are away, rays will play: effects of top predator removal in coral reef ecosystems. *Marine Ecology Progress Series, 641*, 145–157. https://www.int-res.com/abstracts/meps/v641/p145-157/

Shiffman, D. S. (2014, September 3). Sharks aren't always the top of the food chain. *Southern Fried Science.* https://www.southernfriedscience.com/sharks-arent-always-the-top-of-the-food-chain/

Shiffman, D. S. (2019, December 11). Even more amazing shark facts you haven't heard a million times. *Scuba Diving Magazine.* https://www.scubadiving.com/even-more-amazing-shark-facts-you-havent-heard-million-times

Ward, P., & Myers, R. A. (2005). Shifts in open-ocean fish communities coinciding with the commencement of commercial fishing. *Ecology, 86*(4), 835–847. https://esajournals.onlinelibrary.wiley.com/doi/abs/10.1890/03-0746

Chapter 4: What Are the Threats to Sharks and How Threatened Are They?

Associated Press (2019, September 16). Fisherman gets 10 days for dragging live shark behind boat. *Associated Press.* https://www.sun-sentinel.com/news/florida/fl-ne-shark-dragging-sentencing-20190916-wbnmhdna4zdrposafomctlr6nq-story.html

Bale, R. (2017, August 15). Shark fishing forum reveals destructive practices despite good intentions. *National Geographic.* https://www.nationalgeographic.com/animals/article/wildlife-watch-land-based-shark-fishing-florida-study

Bangley, C. W., Paramore, L., Shiffman, D. S., & Rulifson, R. A. (2018). Increased abundance and nursery habitat use of the bull shark (Carcharhinus leucas) in response to a changing environment in a warm-temperate estuary. *Scientific Reports, 8*(1), 1–10. https://www.nature.com/articles/s41598-018-24510-z/?sf187150775=1

Beerkircher, L. R., Cortes, E., & Shivji, M. (2002). Characteristics of shark bycatch observed on pelagic longlines off the southeastern United States, 1992–2000. *Marine Fisheries Review, 64*(4), 40–49. https://spo.nmfs.noaa.gov/sites/default/files/pdf-content/MFR/mfr644/mfr6443.pdf

Black, R. (2013, February 4). Debate continues: Did your seafood feel pain? *National Geographic.* https://www.nationalgeographic.com/animals/article/130208-seafood-pain-debate-crabs-fish-science

Chin, A., Kyne, P. M., Walker, T. I., & McAuley, R. B. (2010). An integrated risk assessment for climate change: analysing the vulnerability of sharks and rays on Australia's Great Barrier Reef. *Global Change Biology, 16*(7), 1936–1953. https://onlinelibrary.wiley.com/doi/abs/10.1111/j.1365-2486.2009.02128.x

Clarke, S. C., McAllister, M. K., Milner-Gulland, E. J., Kirkwood, G. P., Michielsens, C. G. J., Agnew, D. J., Pikitch, E. K., Nakano, H., & Shivji, M. S. (2006). Global estimates of shark catches using trade records from commercial markets. *Ecology Letters, 9*(10), 1115–1126. https://onlinelibrary.wiley.com/doi/abs/10.1111/j.1461-0248.2006.00968.x

Coleman, F. C., Figueira, W. F., Ueland, J. S., & Crowder, L. B. (2004). The impact of United States recreational fisheries on marine fish populations. *Science, 305*(5692), 1958–1960. https://science.sciencemag.org/content/305/5692/1958.abstract

Cooke, S. J., & Cowx, I. G. (2004). The role of recreational fishing in global fish crises. *BioScience, 54*(9), 857–859. https://academic.oup.com/bioscience/article-abstract/54/9/857/252977

Dent, F. and Clarke, S. (2015). State of global markets for shark products. FAO Fisheries and Aquaculture Technical Paper. https://www.proquest.com/openview/3b5c990099f5140bc44e63e7e691e271/1?pq-origsite=gscholar&cbl=237320

Germanov, E. S., Marshall, A. D., Bejder, L., Fossi, M. C., & Loneragan, N. R. (2018). Microplastics: No small problem for filter-feeding megafauna. *Trends in Ecology & Evolution, 33*(4), 227–232. https://www.sciencedirect.com/science/article/pii/S0169534718300090

Newitz, A. (2012, May 21). Lies you've been told about the Pacific Garbage Patch. *Gizmodo.* https://gizmodo.com/lies-youve-been-told-about-the-pacific-garbage-patch-5911969

Parton, K. J., Godley, B. J., Santillo, D., Tausif, M., Omeyer, L. C., & Galloway, T. S. (2020). Investigating the presence of microplastics in demersal sharks of the North-East Atlantic. *Scientific Reports, 10*(1), 1–11. https://www.nature.com/articles/s41598-020-68680-1

Potts, W. M., Downey-Breedt, N., Obregon, P., Hyder, K., Bealey, R., & Sauer, W. H. (2020). What constitutes effective governance of recreational fisheries?—A global review. *Fish and Fisheries, 21*(1), 91–103. https://onlinelibrary.wiley.com/doi/abs/10.1111/faf.12417

Shiffman, D. S. (2015, February 26). Trade in shark fins takes a plunge. *Scientific American.* https://www.scientificamerican.com/article/trade-in-shark-fins-takes-a-plunge/

Shiffman, D. S. (2017, June 13). I asked 15 ocean plastic pollution experts about the Ocean Cleanup project. They have concerns. *Southern Fried Science.* https://www.southernfriedscience.com/i-asked-15-ocean-plastic-pollution-experts-about-the-ocean-cleanup-project-and-they-have-concerns/

Shiffman, D. S. (2019, July 26). Shark tagging: myths, misconceptions, and

nonsense. *Scuba Diving Magazine.* https://www.scubadiving.com/shark
-tagging-myths-misconceptions-and-nonsense

Shiffman, D. S. (2019, August 2). Can sharks get cancer? *Sport Diver Magazine.*
https://www.sportdiver.com/can-sharks-get-cancer

Shiffman, D. S. (2020, January 14). What the hell is DC Metro's "climate change
will increase shark bites" ad talking about? *Southern Fried Science.* https://
www.southernfriedscience.com/what-the-hell-is-the-dc-metros-climate
-change-will-increase-shark-bites-ad-talking-about-an-investigation/

Shiffman, D. S. (2020, January 18). Can wildlife tourism solve the shark
conservation crisis? *Scuba Diving Magazine.* https://www.scubadiving.com
/can-wildlife-tourism-solve-shark-conservation-crisis

Shiffman, D. S. (2020, March 16). Fishing for fun has a bigger environmental
impact than we thought. *Revelator News.* https://therevelator.org/recreational
-fishing-environmental-impact/

Shiffman, D. S., Bittick, S. J., Cashion, M. S., Colla, S. R., Coristine, L. E.,
Derrick, D. H., . . . & Dulvy, N. K. (2020). Inaccurate and biased global
media coverage underlies public misunderstanding of shark conservation
threats and solutions. *iScience*, 23(6), 101205. https://www.sciencedirect
.com/science/article/pii/S2589004220303904

Shiffman, D. S., Gallagher, A. J., Wester, J., Macdonald, C. C., Thaler,
A. D., Cooke, S. J., & Hammerschlag, N. (2014). Trophy fishing for
species threatened with extinction: A way forward building on a history of
conservation. *Marine Policy*, 50, 318–322. https://www.sciencedirect.com
/science/article/pii/S0308597X14001754

Shiffman, D. S., & Hammerschlag, N. (2014). An assessment of the scale,
practices, and conservation implications of Florida's charter boat–based
recreational shark fishery. *Fisheries*, 39(9), 395–407. https://www.tandfonline
.com/doi/abs/10.1080/03632415.2014.941439

Shiffman, D. S., Macdonald, C., Ganz, H. Y., & Hammerschlag, N. (2017).
Fishing practices and representations of shark conservation issues among users
of a land-based shark angling online forum. *Fisheries Research, 196*, 13–26.
https://www.sciencedirect.com/science/article/pii/S016578361730214X

Stevens, J. D., Bonfil, R., Dulvy, N. K., & Walker, P. A. (2000). The effects of
fishing on sharks, rays, and chimaeras (chondrichthyans), and the implications
for marine ecosystems. *ICES Journal of Marine Science, 57*(3), 476–494.
https://academic.oup.com/icesjms/article-abstract/57/3/476/635915

Stump, K. L., Crooks, C. J., Fitchett, M. D., Gruber, S. H., & Guttridge, T.
L. (2017). Hunted hunters: an experimental test of the effects of predation
risk on juvenile lemon shark habitat use. *Marine Ecology Progress Series, 574*,
85–95. https://www.int-res.com/abstracts/meps/v574/p85-95/

Wilkinson, A. (2014, August 12). Scientists want end to traditional trophy
fishing of threatened species. *Science.* https://www.sciencemag.org
/news/2014/08/scientists-want-end-traditional-trophy-fishing-threatened
-species

Worm, B., Davis, B., Kettemer, L., Ward-Paige, C. A., Chapman, D., Heithaus, M. R., Kessel, S. T., & Gruber, S. H. (2013). Global catches, exploitation rates, and rebuilding options for sharks. *Marine Policy, 40*, 194–204. https://www.sciencedirect.com/science/article/abs/pii/S0308597X13000055

Chapter 5: How Can We Protect Sharks?
Albert, C., Luque, G. M., & Courchamp, F. (2018). The twenty most charismatic species. *PLOS ONE, 13*(7), e0199149. https://journals.plos.org/plosone/article?id=10.1371/journal.pone.0199149

Araujo, G., Snow, S., So, C. L., Labaja, J., Murray, R., Colucci, A., & Ponzo, A. (2017). Population structure, residency patterns, and movements of whale sharks in Southern Leyte, Philippines: results from dedicated photo-ID and citizen science. *Aquatic Conservation: Marine and Freshwater Ecosystems, 27*(1), 237–252. https://onlinelibrary.wiley.com/doi/abs/10.1002/aqc.2636

Dulvy, N. K., Fowler, S. L., Musick, J. A., Cavanagh, R. D., Kyne, P. M., Harrison, L. R., . . . & White, W. T. (2014). Extinction risk and conservation of the world's sharks and rays. *eLife, 3*, e00590. https://elifesciences.org/articles/590

Dulvy, N. K., Simpfendorfer, C. A., Davidson, L. N., Fordham, S. V., Bräutigam, A., Sant, G., & Welch, D. J. (2017). Challenges and priorities in shark and ray conservation. *Current Biology, 27*(11), R565–R572. https://www.sciencedirect.com/science/article/pii/S0960982217304827

Graham, F., Rynne, P., Estevanez, M., Luo, J., Ault, J. S., & Hammerschlag, N. (2016). Use of marine protected areas and exclusive economic zones in the subtropical western North Atlantic Ocean by large highly mobile sharks. *Diversity and Distributions, 22*(5), 534–546. https://onlinelibrary.wiley.com/doi/abs/10.1111/ddi.12425

Kitchell, J. F., Essington, T. E., Boggs, C. H., Schindler, D. E., & Walters, C. J. (2002). The role of sharks and longline fisheries in a pelagic ecosystem of the central Pacific. *Ecosystems, 5*(2), 202–216. https://link.springer.com/article/10.1007/s10021-001-0065-5

Libralato, S., Christensen, V., & Pauly, D. (2006). A method for identifying keystone species in food web models. *Ecological Modelling, 195*(3-4), 153–171. https://www.sciencedirect.com/science/article/pii/S0304380005006149

Okey, T. A., Banks, S., Born, A. F., Bustamante, R. H., Calvopiña, M., Edgar, G. J., . . . & Wallem, P. (2004). A trophic model of a Galápagos subtidal rocky reef for evaluating fisheries and conservation strategies. *Ecological Modelling, 172*(2-4), 383–401. https://www.sciencedirect.com/science/article/pii/S0304380003003818

Shiffman, D. S. (2020, January 8). To save endangered sharks, you sometimes need to kill a few. *Scientific American.* https://blogs.scientificamerican.com/observations/to-save-endangered-sharks-you-sometimes-need-to-kill-a-few/

Shiffman, D. S. (2020, July 11). What is the IUCN Red List? *Scuba Diving Magazine.* https://www.scubadiving.com/what-is-iucn-red-list

Stein, R. W., Mull, C. G., Kuhn, T. S., Aschliman, N. C., Davidson, L. N. K., Joy, J. B., & Mooers, A. O. (2018). Global priorities for conserving the evolutionary history of sharks, rays and chimaeras. *Nature Ecology & Evolution*, 2(2), 288–298. https://www.nature.com/articles/s4155 9-017-0448-4

Chapter 6: Sustainable Fisheries for Shark Conservation

Afonso, A. S., Hazin, F. H., Carvalho, F., Pacheco, J. C., Hazin, H., Kerstetter, D. W., Murie, D., & Burgess, G. H. (2011). Fishing gear modifications to reduce elasmobranch mortality in pelagic and bottom longline fisheries off Northeast Brazil. *Fisheries Research*, 108(2–3), 336–343.

Bangley, C. (2013, March 27). Of fin to body ratios and smooth dogfish. *Southern Fried Science*. https://www.southernfriedscience.com/of-fin-body -ratios-and-smooth-dogfish/

Chosid, D. M., Pol, M., Szymanski, M., Mirarchi, F., & Mirarchi, A. (2012). Development and observations of a spiny dogfish Squalus acanthias reduction device in a raised footrope silver hake Merluccius bilinearis trawl. *Fisheries Research*, 114, 66–75. https://www.sciencedirect.com/science/article/pii /S0165783611001160

Gilman, E. (2007). *Shark depredation and unwanted bycatch in pelagic longline fisheries: Industry practices and attitudes, and shark avoidance strategies.* Western Pacific Regional Fishery Management Council. https://stg-wedocs.unep.org /xmlui/handle/20.500.11822/13627

Hall, M. A. (1998). An ecological view of the tuna–dolphin problem: Impacts and trade-offs. *Reviews in Fish Biology and Fisheries*, 8(1), 1–34. https:// link.springer.com/article/10.1023/A:1008854816580

Kynoch, R. J., Fryer, R. J., & Neat, F. C. (2015). A simple technical measure to reduce bycatch and discard of skates and sharks in mixed-species bottom-trawl fisheries. *ICES Journal of Marine Science*, 72(6), 1861–1868. https://academic .oup.com/icesjms/article-abstract/72/6/1861/921176

Peterson, C. D., Belcher, C. N., Bethea, D. M., Driggers III, W. B., Frazier, B. S., & Latour, R. J. (2017). Preliminary recovery of coastal sharks in the south-east United States. *Fish and Fisheries*, 18(5), 845–859. https://onlinelibrary.wiley .com/doi/abs/10.1111/faf.12210

Porsmoguer, S. B., Bănaru, D., Boudouresque, C. F., Dekeyser, I., & Almarcha, C. (2015). Hooks equipped with magnets can increase catches of blue shark (Prionace glauca) by longline fishery. *Fisheries Research*, 172, 345–351. https://www.sciencedirect.com/science/article/pii/S0165783615300254

Shiffman, D. S. (2011, January 7). Turtle excluder devices: analysis of resistance to a successful conservation policy. *Southern Fried Science*. https://www. southernfriedscience.com/turtle-excluder-devices-analysis-of-resistance-to-a -successful-conservation-policy/

Shiffman, D. S. (2014, July 7). Scalloped hammerheads become the first shark species on the U.S. endangered species list. *Scientific American*. https://

www.scientificamerican.com/article/scalloped-hammerheads-become-first
-shark-species-on-the-u-s-endangered-species-list/

Shiffman, D. S. (2015, September 24). Sharks can sense electricity, and that might save them from extinction. *Gizmodo*. https://gizmodo.com/sharks-can -sense-electricity-and-that-might-save-them-1732843117

Shiffman, D. S. (2016, September 23). The most important conservation event you've never heard of is about to start. *The Washington Post*. https://www .washingtonpost.com/news/speaking-of-science/wp/2016/09/23/the-most -important-conservation-event-youve-never-heard-of-is-about-to-start/

Shiffman, D. S., & Hammerschlag, N. (2016). Preferred conservation policies of shark researchers. *Conservation Biology*, *30*(4), 805–815. https://conbio .onlinelibrary.wiley.com/doi/abs/10.1111/cobi.12668

Shiffman, D. S., & Hammerschlag, N. (2016). Shark conservation and management policy: a review and primer for non-specialists. *Animal Conservation*, *19*(5), 401–412. https://zslpublications.onlinelibrary.wiley.com /doi/abs/10.1111/acv.12265

Simpfendorfer, C. A., & Dulvy, N. K. (2017). Bright spots of sustainable shark fishing. *Current Biology*, *27*(3), R97–R98. https://www.sciencedirect.com /science/article/pii/S0960982216314646

Thorpe, T., & Frierson, D. (2009). Bycatch mitigation assessment for sharks caught in coastal anchored gillnets. *Fisheries Research*, *98*(1–3), 102–112. https://www.sciencedirect.com/science/article/pii/S0165783609000964

Ward, P., Lawrence, E., Darbyshire, R., & Hindmarsh, S. (2008). Large-scale experiment shows that nylon leaders reduce shark bycatch and benefit pelagic longline fishers. *Fisheries Research*, *90*(1–3), 100–108. https:// www.sciencedirect.com/science/article/pii/S0165783607002512

Ward, P., Myers, R. A., & Blanchard, W. (2004). Fish lost at sea: the effect of soak time on pelagic longline catches. *Fishery Bulletin*, *102*(1), 179–195. http://aquaticcommons.org/id/eprint/15040

Watson, J. T., Essington, T. E., Lennert-Cody, C. E., & Hall, M. A. (2009). Trade-offs in the design of fishery closures: Management of silky shark bycatch in the Eastern Pacific Ocean tuna fishery. *Conservation Biology*, *23*(3), 626–635. https://conbio.onlinelibrary.wiley.com/doi/abs/10.1111/j.1523 -1739.2008.01121.x

Wilcox, C. (2014, July 13). "Endangered"—You keep using that word. I do not think it means what you think it means. *Discover Magazine*. https://www .discovermagazine.com/planet-earth/endangeredyou-keep-using-that-word-i -do-not-think-it-means-what-you-think-it-means

Chapter 7: Fishing and Trade Bans for Shark Conservation

Ali, K., & Sinan, H. (2014). Shark ban in its infancy: Successes, challenges and lessons learned. *Journal of the Marine Biological Association of India*, *56*, 34–40. https://www.bmis-bycatch.org/system/files/zotero_attachment

s/library_1/4FHEESE6%20-%20Journal%20of%20the%20Marine%20
Biological%20Association%20of%20India-vol.56-no.1.pdf#page=34

Bowcott, O. (2020, May 12). Chagos islander's exile is ongoing breach of human
rights. *The Guardian.* https://www.theguardian.com/world/2020/may/12
/chagos-islanders-exile-human-rights-breach-court-of-appeal-told

Da Silva, C., Kerwath, S. E., Attwood, C. G., Thorstad, E. B., Cowley, P. D.,
Økland, F., Wilke, C. G., & Næsje, T. F. (2013). Quantifying the degree of
protection afforded by a no-take marine reserve on an exploited shark. *African
Journal of Marine Science, 35*(1), 57–66. https://www.tandfonline.com/doi/abs
/10.2989/1814232X.2013.769911

Davidson, L. N., & Dulvy, N. K. (2017). Global marine protected areas to
prevent extinctions. *Nature Ecology & Evolution, 1*(2), 1–6. https://
www.nature.com/articles/s41559-016-0040

Dulvy, N. K. (2013). Super-sized MPAs and the marginalization of species
conservation. *Aquatic Conservation: Marine and Freshwater Ecosystems, 23*(3),
357–362. https://onlinelibrary.wiley.com/doi/abs/10.1002/aqc.2358

Dwyer, R. G., Krueck, N. C., Udyawer, V., Heupel, M. R., Chapman, D., Pratt
Jr., H., Garla, R., & Simpfendorfer, C. A. (2020). Individual and population
benefits of marine reserves for reef sharks. *Current Biology, 30*(3), 480–489.
https://www.sciencedirect.com/science/article/pii/S0960982219316008

Edgar, G. J., Stuart-Smith, R. D., Willis, T. J., Kininmonth, S., Baker, S. C.,
Banks, S., . . . & Thomson, R. J. (2014). Global conservation outcomes
depend on marine protected areas with five key features. *Nature, 506*(7487),
216–220. https://www.nature.com/articles/nature13022

Hammerschlag, N., Gallagher, A. J., Wester, J., Luo, J., & Ault, J. S. (2012).
Don't bite the hand that feeds: Assessing ecological impacts of
provisioning ecotourism on an apex marine predator. *Functional Ecology,
26*(3), 567–576.

Lester, S. E., Halpern, S. B., Grorud-Colvert, K., Lubchenco, J., Ruttenberg, B.
I., Gaines, S. D., Airamé, S., & Warner, R. R. (2009). Biological effects within
no-take marine reserves: a global synthesis. *Marine Ecology Progress Series, 384,*
33–46. https://www.int-res.com/abstracts/meps/v384/p33–46

Macpherson, R. (2018, March 23). Embracing yes/also: Marine protected areas
are not an either/or proposition. *Deep Sea News.* https://www.deepseanews.
com/2018/03/embracing-yes-also-marine-protected-areas-are-not-an-either
-or-proposition/

Rigby, C. L., Simpfendorfer, C. A., & Cornish, A. (2019). A practical guide to
effective design and management of MPAs for sharks and rays. *WWF Report.*
https://sharks.panda.org/images/PDF/WWF_MPA
_Guide2019.pdf

Rocha, L. A. (2018, March 20). Bigger is not better for ocean conservation. *The
New York Times.* https://www.nytimes.com/2018/03/20/opinion
/environment-ocean-conservation.html

Roy, E. A. (2017, July 15). "Quite odd": coral and fish thrive at Bikini Atoll 70 years after nuclear tests. *The Guardian.* https://www.theguardian.com/world/2017/jul/15/quite-odd-coral-and-fish-thrive-on-bikini-atoll-70-years-after-nuclear-tests

Shiffman, D. S. (2019, June 6). How to protect sharks from overfishing. *Revelator News.* https://therevelator.org/protect-sharks-overfishing/

Shiffman, D. S. (2019, November 21). The recipe for a successful MPA. *Scuba Diving Magazine.* https://www.scubadiving.com/successful-marine-protected-areas-mpas-analysis

Shiffman, D. S. (2020, October 4). An ambitious strategy to preserve biodiversity. *Scientific American.* https://www.scientificamerican.com/article/an-ambitious-strategy-to-preserve-biodiversity/

Shiffman, D. S., & Hueter, R. E. (2017). A United States shark fin ban would undermine sustainable shark fisheries. *Marine Policy, 85,* 138–140. https://www.sciencedirect.com/science/article/pii/S0308597X17304384

Silver, N. (2010, April 19). Double down by the numbers: The unhealthiest sandwich ever? *FiveThirtyEight.* https://fivethirtyeight.com/features/double-down-by-numbers-unhealthiest/

Vianna, G. M., Meekan, M. G., Ruppert, J. L., Bornovski, T. H., & Meeuwig, J. J. (2016). Indicators of fishing mortality on reef-shark populations in the world's first shark sanctuary: the need for surveillance and enforcement. *Coral Reefs, 35*(3), 973–977. https://link.springer.com/article/10.1007/s00338-016-1437-9?platform=hootsuite

White, T. D., Carlisle, A. B., Kroodsma, D. A., Block, B. A., Casagrandi, R., De Leo, G. A., Gatto, M., Micheli, F., & McCauley, D. J. (2017). Assessing the effectiveness of a large marine protected area for reef shark conservation. *Biological Conservation, 207,* 64–71. https://www.sciencedirect.com/science/article/abs/pii/S0006320717300678

Chapter 8: How Are Scientists Helping Sharks?

Burgess, G. H., Beerkircher, L. R., Cailliet, G. M., Carlson, J. K., Cortés, E., Goldman, K. J., . . . & Simpfendorfer, C. A. (2005). Is the collapse of shark populations in the Northwest Atlantic Ocean and Gulf of Mexico real? *Fisheries, 30*(10), 19–26.

Macdonald, C. (2020, August 10). The dark side of being a female shark researcher. *Scientific American.* https://www.scientificamerican.com/article/the-dark-side-of-being-a-female-shark-researcher/

Shiffman, D. S. (2014, January 13). Shark species thought to be extinct found in fish market. *Scientific American.* https://www.scientificamerican.com/article/shark-species-thought-to-be-extinct-found-in-fish-market/

Shiffman, D. S. (2014, August 11). Experts: Shark Week's "zombie sharks" harasses animals. *Earth Touch News.* https://www.earthtouchnews.com/oceans/sharks/experts-shark-weeks-zombie-sharks-harasses-animals/

Shiffman, D. S. (2019, August 26). My shark friend on Twitter died, now what? *Hakai Magazine*. https://www.hakaimagazine.com/news/my-shark-friend-on -twitter-died-now-what/

Shiffman, D. S. (2019, September 4). Scientific research in the fish market. *Hakai Magazine*. https://www.hakaimagazine.com/news/scientific-research-in -the-fish-market/

Shiffman, D. S. (2020, February 3). How can you tell how old a shark is? *Sport Diver Magazine*. https://www.sportdiver.com/how-can-you-tell-how-old -shark-is

Shiffman, D. S. (2020, April 23). How can you become a marine biologist too? *Sport Diver magazine*. https://www.sportdiver.com/how-can-you-become -marine-biologist-too

Shiffman, D. S. (2021, March 11). What exactly is marine conservation biology? *Ocean Conservancy* blog. https://oceanconservancy.org/blog/2021/03/11 /exactly-marine-conservation-biology/

Shiffman, D. S., Ajemian, M. J., Carrier, J. C., Daly-Engel, T. S., Davis, M. M., Dulvy, N. K., . . . & Wyffels, J. T. (2020). Trends in chondrichthyan research: an analysis of three decades of conference abstracts. *Copeia*, *108*(1), 122–131. https://meridian.allenpress.com/copeia/article-abstract/108/1/122/434978

Sofia, M. (2021, July 19). Building a shark science community for women of color. *National Public Radio Short Wave*. https://www.npr. org/2021/07/14/1016089180/building-a-shark-science-community-for -women-of-color

Steel, B., List, P., Lach, D., & Shindler, B. (2004). The role of scientists in the environmental policy process: A case study from the American west. *Environmental Science & Policy*, *7*(1), 1–13. https://www.sciencedirect.com /science/article/pii/S1462901103001400

INDEX

Page numbers in *italic* refer to illustrations.